AF557751

BASICS OF GENETICS

BASICS OF GENETICS

SWARNAPRABHA PRADHAN

ANMOL PUBLICATIONS PVT. LTD.
NEW DELHI - 110 002 (INDIA)

ANMOL PUBLICATIONS PVT. LTD.

H.O.: 4374/4B, Ansari Road, Darya Ganj,
New Delhi-110 002 (India)
Ph.: 23278000, 23261597

B.O.: No. 1015, Ist Main Road, BSK IIIrd Stage
IIIrd Phase, IIIrd Block
Bangalore - 560 085 (India)
Visit us at: www.anmolpublications.com

Basics of Genetics

ISBN 978-81-261-3499-1

PRINTED IN INDIA

Printed at Mehra Offset Press, Delhi.

Contents

Preface

Modern biology is rooted in an understanding of the molecules within cells and of the interactions between cells that allow construction of multicellular organisms. The more we learn about the structure, function, and development of different organisms, the more we recognize that all life processes exhibit remarkable similarities. Basics of Genetics concentrates on the macromolecules and reactions studied by biochemists, the processes described by scientists and the gene control pathways identified by molecular biologists and geneticists.

All the concepts of Genetics continue to be derived from experiments, and powerful experimental tools that allow the study of living cells and organisms at higher and higher levels of resolution are being developed constantly. In this edition, we address the current state of cell biology and look forward to what further exploration will uncover in the twenty-first century.

Author

Chapter 1

Introduction

Genetics is the study of the function and behaviour of genes. Genes are bits of biochemical instructions found inside the cells of every organism from bacteria to humans. Offspring receive a mixture of genetic information from both parents. This process contributes to the great variation of traits that we see in nature, such as the colour of a flower's petals, the markings on a butterfly's wings, or such human behavioral traits as personality or musical talent. Geneticists seek to understand how the information encoded in genes is used and controlled by cells and how it is transmitted from one generation to the next. Geneticists also study how tiny variations in genes can disrupt an organism's development or cause disease. Increasingly, modern genetics involves genetic engineering, a technique used by scientists to manipulate genes. Genetic engineering has produced many advances in medicine and industry, but the potential for abuse of this technique has also presented society with many ethical and legal controversies.

Genetic information is encoded and transmitted from generation to generation in deoxyribonucleic acid (DNA). DNA is a coiled molecule organized into structures called chromosomes within cells. Segments along the length of a DNA molecule form genes. Genes direct the synthesis of proteins, the molecular laborers that carry out all lifesupporting activities in the cell. Although all humans share the same set of genes, individuals can inherit different forms of a given gene, making each person genetically unique.

Since the earliest days of plant and animal domestication, around 10,000 years ago, humans have understood that characteristic traits of parents could be transmitted to their offspring. The first to speculate about how this process worked were Greek scholars around the 4th century bc, who promoted theories based on conjecture or superstition. Some of these theories remained in favour for several centuries. The scientific study of genetics did not begin until the late 19th century. In experiments with garden peas, Austrian monk Gregor Mendel described the patterns of inheritance, observing that traits were inherited as separate units. These units are now known as genes. Mendel's work formed the foundation for later scientific achievements that heralded the era of modern genetics.

Most of the discoveries about genes and DNA happened after Charles Darwin 1859 famous book on The Origin of Species by Means of Natural Selection, or the preservation of favoured races in the struggle for life.

Genetics is the science of heredity and variation in living organisms. Knowledge that desired characteristics were inherited has been implicitly used since prehistoric times for improving crop plants and animals through selective breeding. However, the modern science of genetics, which seeks to understand the mechanisms of inheritance, only began with the work of Gregor Mendel in the mid-1800s.

Mendel observed that inheritance is fundamentally a discrete process with specific traits that are inherited in an independant manner. These basic units of inheritance is now known as "genes". In the cells of organisms, genes exist physically in the structure of the molecule DNA and the information genes contain is used to create and control the components of cells. Although genetics plays a large role in determining the appearance and behaviour of organisms, it is the interaction of genetics with the environment an organism experiences that determines the ultimate outcome. For example, while genes play a role in determining a person's height, the nutrition and health that person experiences in childhood also have a large effect.

Classical genetics consists of the techniques and methodologies of genetics that predate the advent of molecular biology. A key discovery of classical genetics in eukaryotes was genetic linkage. The observation that some genes do not segregate independently at meiosis, broke the laws of Mendelian inheritance, and provided science with a way to map characteristics to a location on the chromosomes. Linkage maps are still used today, especially in breeding for plant improvement.

After the discovery of the genetic code and such tools of cloning as restriction enzymes, the avenues of investigation open to geneticists were greatly broadened. Some classical genetic ideas have been supplanted with the mechanistic understanding brought by molecular discoveries, but many remain intact and in use. Classical genetics is often contrasted with reverse genetics, and aspects of molecular biology are sometimes referred to as molecular genetics.

GENE

A gene is a segment of nucleic acid that contains the information necessary to produce a functional RNA product in a controlled manner. They contain regulatory regions dictating under what conditions this product is made, transcribed regions dictating the sequence of the RNA product, and/or other functional sequence regions. Genes interact with each other and the environment to determine the physical development and behaviour of organisms, they are units of inheritance.

In cells, genes consist of a long strand of DNA that contains a promoter, which controls the activity of a gene, and a coding sequence, which determines what the gene produces. When a gene is active, the coding sequence is copied in a process called transcription, producing an RNA copy of the gene's information. This RNA can then direct the synthesis of proteins via the genetic code. However, RNAs can also be used directly, for example as part of the ribosome.These molecules resulting from gene expression, whether RNA or protein, are known as gene products.

Most genes contain non-coding regions that do not code for the gene products, but regulate gene expression. The genes of eukaryotic organisms can contain non-coding regions called introns that are removed from the messenger RNA in a process known as splicing. The regions that actually encode the gene product, which can be much smaller than the introns, are known as exons. One single gene can lead to the synthesis of multiple proteins through the different arrangements of exons produced by alternative splicings.

The total complement of genes in an organism or cell is known as its genome. The genome size of an organism is loosely dependent on its complexity; prokaryotes such as bacteria and archaea have generally smaller genomes, both in number of base pairs and number of genes, than even single-celled eukaryotes. However, the largest known genome belongs to the single-celled amoeba Amoeba dubia, with over 6 billion base pairs. The estimated number of genes in the human genome has been repeatedly revised downward since the completion of the Human Genome Project; current estimates place the human genome at just under 3 billion base pairs and about 20,000-25,000 genes.. A recent Science article gives a final number of 20,488, with perhaps 100 more yet to be discovered . The gene density of a genome is a measure of the number of genes per million base pairs (called a megabase, Mb); prokaryotic genomes have much higher gene densities than eukaryotes. The gene density of the human genome is roughly 12-15 genes/Mb.

Chapter 2

The Principles of Genetics

How a living organism is built and function determined and govern by genes. By understanding how genes work:

- It would give the possibility to cure genetic illnesses.
- It would give the possibility to determine the organism's features and behaviors.
- It would give the possibility to create a new organism.

The relationships between genes and features are very complex. It may not be possible to change one and only one feature of an organism by changing one or set of genes(based on current knowledge). A feature may be changed, but probably not alone, only with other changes. The other changes could be small and/or ignorable however. Some changes are gradual, while changes in gene expression can result in rapid transformations in the physiological state of an organism.

DOMINANT AND RECESSIVE GENES

DOMINANT AND RECESSIVE CHARACTERS

Substitution has been made in the case of the two kinds of daughters, with the same effect. Wherever an X-chromosome from the original father (white) is substituted for one from the original mother (black), the animal has bar-eyes. Where- ever an X-chromosome from the mother (black) is substituted for one from the father (white), the individual has normal eyes instead of bar-eyes.

It is certain therefore that bar-eye is an abnormality resulting from a defect in certain X-chromosomes. It is clear that X-chromosomes can become defective, and that this change in the X-chromosome causes the individual to develop in an abnormal way, so that it possesses a bodily abnormalityÿÿin this case, an abnormality of the eye. This abnormality appears, as above seen, in any individual that has the defective X-chromosome, whether it also has a normal chromosome or not.

DOMINANT AND RECESSIVE CHARACTERS RESULTING FROM SUBSTITUTION OF CHROMOSOMES

The daughters produced in the experiment of contain X-chromosomes from two different sources one from the normal-eyed mother, one from the bar-eyed father.The one from the mother gives when by itself a normal eye, the one from the father a bar-eye. As the daughters have bar-eyes, it is evident that the effect of the chromosome from the father prevails over the effect of that from the mother. It is the custom to call the character that thus prevails and is manifested, when two chromosomes of different tendency are present, a dominant character. Thus bar-eye is dominant, as compared with normal eye.

Other X-chromosomes, in other individuals of Drosophila, are found to produce still other effects. Some are defective or abnormal, but instead of producing bar-eyes they produce other abnormalities. And most of them behave in a way that differs from the behaviour of bar-eye. An example will make clear this behaviour.

In certain cases X-chromosomes produce eyes that are white instead of red. Normally in the fruit-fly the eyes are red. But sometimes there are found individuals whose eyes are white. When these are mated together, they produce offspring all of whose eyes are white. So the white eyes are hereditary.

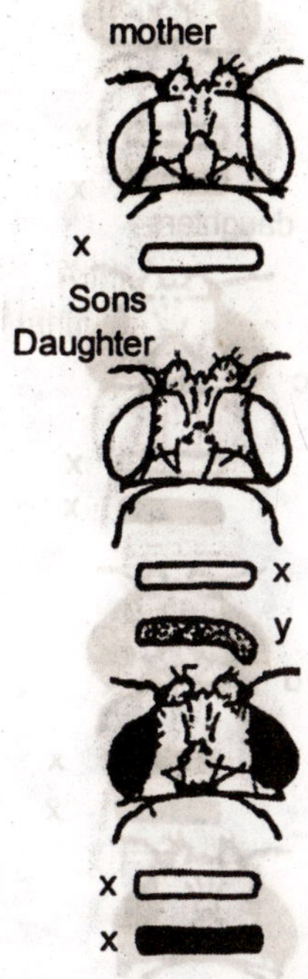

Fig. Inheritance of a recessive characteristic that depends on a particular type of X-chromosome: white eyes. The mother has white eyes, the father red eyes (represented in black]. The X-chromosomes of each individual are represented below the Fig of the head, the mothers chromosomes in outline, the father's in black. The daughters, receiving one X-chromosome from each parent, have red eyes, like the father. The sons, receiving an X-chromosome from the mother only, have white eyes, like the mother. The white eyes appear wherever the maternal type ofX (outline) is the only type ofX that is present.

But what happens if one mates together red-eyed and white-eyed parents? What will be the programmes of the eyes in the offspring? Mate first a white-eyed mother with a red-eyed father.

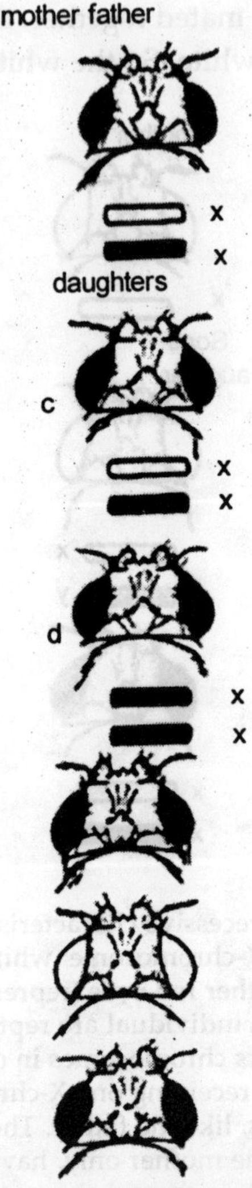

Fig.Course of inheritance of a recessive character (white eyes) that is dependent on a defective X. The mother has one defective X-chromosome (shown in outline), derived from the white-eyed mother of Fig. ii, her other X-chromosome being normal; she therefore has normal red eyes. The father has a normal X-chromosome (black), and normal red eyes. Of the four types of offspring shown (a, b, c, d), only one (c), representing half the sons, have the white eyes, this being the consequence of their receiving the mother's defective X-chromosome. All the daughters (a and b) and the other half of the sons (d) have normal redeyes, since they contain at least one of the normal X-chromosomes.

In this case the Fig represents the mother's X-chromosome in white (since these are the abnormal individuals), while the father's are shown in black.Thus the descendants of the two kinds can be traced in later generations.The sons get an X-chromosome from the mother only. And they all have white eyes, like the mother. The daughters get one X from the mother, one from the father. And they all have red eyes, like the father.

So the white eyes go with the mother's X-chromosome. But if one of the father's X-chromosomes is likewise present, then the white eyes are not produced; such daughters have red eyes. So the red eyes are dominant. The white eyes, since they do not appear if a normal X-chromosome from the father is present along with the abnormal one from the mother, are said to be recessive. White eyes appear only in individuals in which the abnormal X-chromosomes are the only kind of Xpresent.

All this can be tested in a great number of ways. Mate, for example, one of these daughters, having the two kinds of chro-mosomes, with a normal red-eyed male, having a normal (black) X. Now half of the offspring get from the mother a defective X (white), the other half get from her a normal X (black). This is true both for the daughters and the sons. But the daughters get also a normal X from the redeyed father. So they all have normal red eyes. But the sons do not get an X from the father; the only X they have is the one from the mother. And those that get the normal X from the mother have red eyes, while those that get the defective X have white eyes. That is, there is now produced a family in which all the daughters have red eyes, while half the sons have red eyes and the other half have white eyes. In a family of two hundred, there are a hundred red-eyed daughters, fifty red-eyed sons, and fifty white-eyed sons.The white eyes are present in all cases where the efective chromosome from the grandfather is the only kind present. There are many ways to test this, and it is always verified. For example, if we cross one of the red-eyed mothers that has one defective X-chromosome, with a white-eyed father that also has the defective X-chromosome then we find that half the sons, and also half the daughters, have white eyes, the other half red eyes. And this is exactly the result given if the white eyes appear only where the defective X is the only kind present.Since such characters appear in all males that have the defective X (since they have no other), while in females they appear only if both the X's are defective, they are much commoner in males than in females, and are, hence, often called sex-linked characters. But they appear in females as well as in males, provided that both the X's are defective. Many Types of X-Chromosomes exist.Thus it is clear, up to this point, that there are among the different individuals of the fruit-fly at least three kinds of X-chromosomes, with different effects. Some are normal, producing normal red eyes. Others have become defective in such a way that they produce bar-eyes. Others have become defective in such a way that they produce white eyes.These are not the only kinds of changed X-chromosomes.

Many different sorts of peculiarities have been found that thus follow the distribution of a particular type of X-chromosome, being manifested in the body of the individual when that type of X is the only type present. Up to 1925, a hundred and thirty diverse types of X-chromosomes had been detected scattered among the different individuals in the fruit-fly each type giving rise to some special peculiarity. Many are known also in other organisms. In man a considerable number are known, giving rise mainly to defects or diseases.Among the characteristics dependent on changed X-chromosomes in man are some types of colourblindness, also a blood disease known as haemophilia. They follow exactly the distribution of a particular type of X-chromosome. This, ofcourse, shows that in man too the male has but a single X chromosome, while the female has two, for this is the only way in which that method of inheritance can be brought about.

Thus it is established that when one X-chromosome is sub stituted for another, this commonly makes a difference in the characteristics of the individuals. Since there are many types of X-chromosomes distributed among the individuals of a single species, such substitutions can be made in many different ways, giving many different types of results. In the fruitfly, some X-chromosomes produce red eyes, others produce white eyes, others buff coloured, or eosin coloured, or various other coloured eyes. Some X-chromosomes produce long wings, others short wings and so one could list a hundred or more diverse kinds of effects of diverse X-chromosomes. Similar long lists could be made for certain other organisms, and a considerable number of diverse types of X-chromosomes are known even in man.The other chromosomes present somewhat more complex conditions than do the X-chromosomes, because they are always in pairs, while in some individuals (males) X is single. This makes detection of the effects of substituting one chromosome for another more difficult. But the same type of experimentation illustrated above for the X-chromosomes demonstrates that the other chromosomes also influence the characteristics of organisms. These matters are taken up in detail in later chapters. It is clear therefore that the chromosomes are indeed 'materials of heredity'. That is, they influence the characteristics of the individuals that contain them, for these characteristics are altered when one chromosome is substituted for another. We turn next to a study of these materials.

The relations between the materials of heredity and the characteristics of the organism have been far more thoroughly investigated in the fruit-fly, Drosophila melanogaster, than in any other organism, so that much of the detailed knowledge of these matters is based upon this animal. Most of the fundamental work on Drosophila has been done by T. H. Morgan and his associates,G. B. Bridges and A. H. Sturtevant, though in recent years many investigators in all parts of the world have been at work on the genetics of this organism, adding greatly to our knowledge.

When Mendel studied peas, one of the phenotypes showed complete dominance over the other one. If we look at pea height, and denote the gene for short as s and the gene for tall as S, then as every plant has two sets chromosomes each has two genes at this locus.So there are three possibilities: SS, Ss, ss (order doesn't matter) In this case S was fully dominant over s, so Ss individuals were phenotypically identical to SS individuals. Only ss pea plants were short. The S gene would be said to be Dominant.While the s gene is said to be Recessive.

MENDELIAN INHERITENCE

INTRODUCTION

Gregor Johann Mendel was a monk in the Augustinian Monastery in the Brunn, Czech Republic. In 1854 he began the experiments which started modern genetics. His work with garden peas, Pisum sativum, was vital to our understanding of inheritance. He is known as the Father of Genetics.

MENDEL'S EXPERIMENT

Mendel reasoned an organism for genetic experiments should have:

1. A number of different traits that can be studied
2. Plant should be self-fertilizing and have a flower structure that limits accidental contact
3. Offspring of self-fertilized plants should be fully fertile.

Mendel's experimental organism was a common garden pea (Pisum sativum), which has a flower that lends itself to self-pollination. The male parts of the flower are termed the anthers. They produce pollen, which contains the male gametes (sperm). The female parts of the flower are the stigma, style, and ovary. The egg (female gamete) is produced in the ovary. The process of pollination (the transfer of pollen from anther to stigma) occurs prior to the opening of the pea flower. The pollen grain grows a pollen tube which allows the sperm to travel through the stigma and style, eventually reaching the ovary. The ripened ovary wall becomes the fruit (in this case the pea pod). Most flowers allow cross-pollination, which can be difficult to deal with in genetic studies if the male parent plant is not known. Since pea plants are selfpollinators, the genetics of the parent can be more easily understood. Peas are also self-compatible, allowing self-fertilized embryos to develop as readily as out-fertilized embryos. Mendel tested all 34 varieties of peas available to him through seed dealers. The garden peas were planted and studied for eight years. Each character studied had two distinct forms, such as tall or short plant height, or smooth or wrinkled seeds. Mendel's experiments used some 28,000 pea plants.

Mendel's first step was breeding pure breeding strains of peas. The traits he studied included:

- Pea programmes
- Height
- And whether the Peas were wrinkled or smooth.

Mendel crossed the pure breeding Parental Generation (designated P). He found that the first generation (F1) was exclusively phenotypically one of the parental types. Mendel then crossed his F1 generation with itself. He found that the F2 generation showed a surprising trait, three quarters were like the F1 generation, while the remaining quarter were like the other Parents.From this Mendel realised that there were two versions of each loci, one of which expressed dominance over the other. He called this Biparticulate Inheritance. If a gene was following this 3:1 pattern it was said to be segregating normally. By looking at multiple genes, Mendel showed that they were not linked to each other and that each loci he studied had no influence over the others. He called this Independent assortment.

By studying cases where the Mendelian laws we can also learn a lot. For instance, if a gene isn't segragating normally it may be sex linked. If two genes aren't Assorting Independantly they're probably on the same chromosome.

Mendel's laws are the first step to understanding Genetics, they lay down the basic concept of inheritance.

Mendel's contribution was unique because of his methodical approach to a definite problem, use of clear-cut variables and application of mathematics (statistics) to the problem. Gregor Using pea plants and statistical methods, Mendel was able to demonstrate that traits were passed from each parent to their offspring through the inheritance of genes.

FEATURES OF INHERITANCE

Mendelian inheritance (or Mendelian genetics or Mendelism) is a set of primary tenets relating to the transmission of hereditary characteristics from parent organisms to their children; it underlies much of genetics. They were initially derived from the work of Gregor Mendel published in 1865 and 1866 which was "re-discovered" in 1900, and were initially very controversial. When they were integrated with the chromosome theory of inheritance by Thomas Hunt Morgan in 1915, they became the core of classical genetics.

Mendel's work showed:

1. Each parent contributes one factor of each trait shown in offspring.
2. The two members of each pair of factors segregate from each other during gamete formation.
3. The blending theory of inheritance was discounted.
4. Males and females contribute equally to the traits in their offspring.
5. Acquired traits are not inherited.

EXCEPTIONS TO SIMPLE INHERITANCE

Since Mendel's time, our knowledge of the mechanisms of genetic inheritance has grown immensely. For instance, it is now understood that inheriting one allele can, at times, increase the chance of inheriting another or can affect how and when a trait is expressed in an individual's phenotype. Likewise, there are degrees of dominance and recessiveness with some traits. The simple rules of Mendelian inheritance do not apply in these and other exceptions.

Polygenic Traits

Some traits are determined by the combined effect of more than one pair of genes. These are referred to as polygenic or continuous traits. An example of this is human stature. The combined size of all of the body parts from head to foot determines the height of an individual. There is an additive effect. The sizes of all of these body parts are, in turn, determined by numerous genes. Human skin, hair, and eye colour are also polygenic traits because they are influenced by more than one allele at different loci. The result is the perception of continuous gradation in the expression of these traits.

Intermediate Expression

Apparent blending can occur in the phenotype when there is incomplete dominance resulting in an intermediate expression of a trait in heterozygous individuals. For instance, in primroses, snapdragons, and four-o'clocks, red or white flowers are homozygous while pink ones are heterozygous. The pink flowers result because the single "red" allele is unable to code for the production of enough red pigment to make the petals dark red.

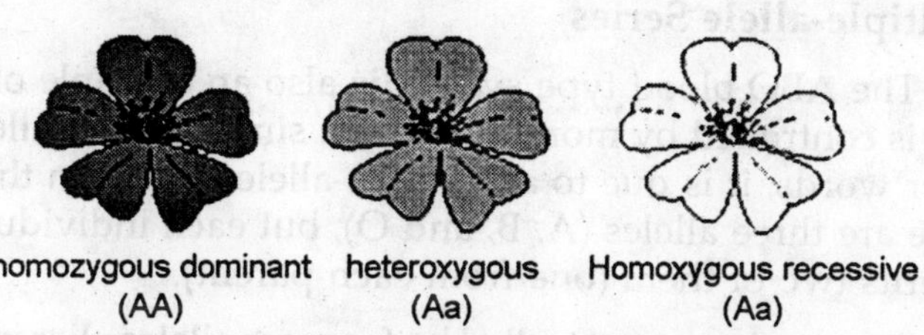

Another example of an intermediate expression may be the pitch of human male voices. The lowest and highest pitches apparently are found in men who are homozygous for this trait (AA and aa), while the intermediate range baritones are heterozygous (Aa). The child-killer disease known as Tay-Sachs is also characterized by incomplete dominance. Heterozygous individuals are genetically programmed to produce only 40-60% of the normal amount of an enzyme that prevents the disease.

Fortunately for Mendel, the pea plant traits that he studied were controlled by genes that do not exhibit an intermediate expression in the phenotype. Otherwise, he probably would not have discovered the basic rules of genetic inheritance.

Codominance

For some traits, two alleles can be codominant. That is to say, both are expressed in heterozygous individuals. An example of this is people who have an AB blood type for the ABO blood system. When they are tested, these individuals actually have the characteristics of both type A and type B blood. Their phenotype is not intermediate between the two.

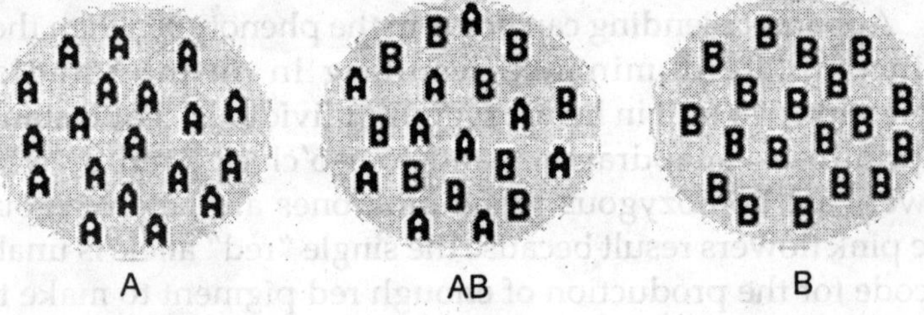

Multiple-allele Series

The ABO blood type system is also an example of a trait that is controlled by more than just a single pair of alleles. In other words, it is due to a multiple-allele series. In this case, there are three alleles (A, B, and O), but each individual only inherits two of them (one from each parent).

Some traits are controlled by far more alleles. For instance, the human HLA system, which is responsible for identifying and rejecting foreign tissue in our bodies, can have at least 30,000,000 different genotypes. It is the HLA system which causes the rejection of organ transplants. The more we learn about human genetics the more it becomes clear that multiple-allele series are very common. In fact, it now appears that they are more common than simple two allele ones.

MODIFYING AND REGULATOR GENES

There are two classes of genes that can have an effect on how other genes function. They are called modifying genes and regulator genes.

Modifying genes alter how certain other genes are expressed in the phenotype. For instance, there is a dominant cataract gene which will produce varying degrees of vision impairment depending on the presence of a specific allele for a companion modifying gene. However, cataracts also can be promoted by diabetes and common environmental factors such as excessive ultraviolet radiation.

Regulator genes can either initiate or block the expression of other genes. They control the production of a variety of chemicals in plants and animals. For instance, the time of production of specific proteins that will be new structural parts of our bodies can be controlled by such regulator genes. Shortly after conception, regulator genes work as master switches orchestrating the timely development of our body parts. They are also responsible for changes that occur in our bodies as we grow older. In other words, they control the maturation and aging processes. Regulator genes that are involved in subdividing an embryo into what will become the major body parts of an individual are also referred to as homeotic or homeobox genes. They are responsible for setting generalized cells on the path to become a head, torso, arms, legs, etc.

Incomplete Penetrance

Some genes are incompletely penetrant. That is to say, their effect does not normally occur unless certain environmental factors are present. For example, you may inherit a gene for diabetes but never get the disease unless you become greatly overweight, persistently stressed psychologically, or do not get enough sleep on a regular basis. Similarly, the gene that causes the chronic disease known as multiple sclerosis may be triggered by the Epstein-Barr virus.

Sex Related Genetic Effects

There are three categories of genes that may have different effects depending on an individual's gender. These are referred to as:

1. Sex-limited genes
2. Sex-controlled genes
3. Genome imprinting

Sex-limited genes are ones that are inherited by both men and women but are normally only expressed in the phenotype of one of them. The heavy male beard is an example. While women have facial hair it is most often very fine and comparatively sparse.

In contrast, sex-controlled genes are expressed in both sexes but differently. An example of this is gout, a disease that causes painfully inflamed joints. If the gene is present, men are nearly eight times more likely than women to have severe symptoms.

Some genes are known to have a different effect depending on the gender of the parent from whom they are inherited. This phenomenon is referred to as genome imprinting or genetic imprinting. Apparently, diabetes, psoriasis, and some rare genetically inherited diseases, such as a form of mental retardation known as Angelman syndrome, can follow this inheritance pattern.

Pleiotropy

A single gene may be responsible for a variety of traits. This is called pleiotropy. The complex of symptoms that are collectively referred to as sickle-cell trait, or sickle-cell anemia, is an example. A single gene results in irregularly shaped red blood cells that painfully block blood vessels, cause poor overall physical development, as well as related heart, lung, kidney, and eye problems. Another pleiotropic trait is albinism. The gene for this trait not only results in a deficiency of skin, hair, and eye pigmentation but also causes defects in vision.

Stuttering Alleles

Lastly, it is now known that some genetically inherited diseases have more severe symptoms each succeeding generation due to segments of the defective genes being doubled in their transmission to children (as illustrated below). These are referred to as stuttering alleles or unstable alleles. Examples of this phenomenon are Huntington's disease, fragile-X syndrome, and the myotonic form of muscular dystrophy.

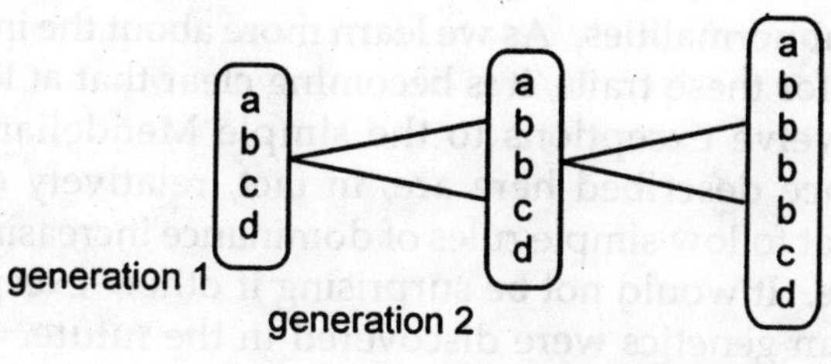

Fig. Unstable allele doubling each generation

Mendel believed that all units of inheritance are passed on to offspring unchanged. Unstable alleles are an important exception to this rule.

Environmental Influences

The phenotype of an individual is not only the result of inheriting a particular set of parental genes.The specific environmental characteristics of the uterus in which a fertilized egg is implanted and the health of the mother can have major impacts on the phenotype of the future child. For instance, oxygen deprivation or inappropriate hormone levels can cause lifelong, devastating effects. Likewise, accidents, poor nutrition, and other environmental influences throughout life can alter an individual's phenotype.

Geneticists study identical or monozygotic twins to determine which traits are inherited and which ones were acquired following conception. Since monozygotic twins come from the same zygote, they are essentially identical in their genetic makeup. If there are any differences in their phenotypes, the environment is virtually always responsible. Such differences show up in basic capabilities such as handedness, which had been assumed to be entirely genetically determined. In rare instances, one monozygotic twin will be clearly right-handed while the other will be left-handed. This suggests that there may be both genetic and environmental influences in the development of this trait.

Summary

Researchers have identified over 15,000 genetically inherited human traits. More than 5,000 of them are diseases or other abnormalities. As we learn more about the inheritance patterns for these traits, it is becoming clear that at least some of the twelve exceptions to the simple Mendelian rules of inheritance described here are, in fact, relatively common. Genes that follow simple rules of dominance increasingly seem to be rare. It would not be surprising if other "exceptions" to Mendelian genetics were discovered in the future.

MENDEL'S LAWS

Law of Segregation

Mendel studied the inheritance of seed shape first.A cross involving only one trait is referred to as a monohybrid cross. Mendel crossed pure-breeding (also referred to as true-breeding) smooth-seeded plants with a variety that had always produced wrinkled seeds (60 fertilizations on 15 plants). All resulting seeds were smooth. The following year, Mendel planted these seeds and allowed them to self-fertilize. He recovered 7324 seeds: 5474 smooth and 1850 wrinkled. To help with record keeping, generations were labeled and numbered. The parental generation is denoted as the P1 generation. The offspring of the P1 generation are the F1 generation (first filial). The selffertilizing F1 generation produced the F2 generation (second fil˙ ˙˙

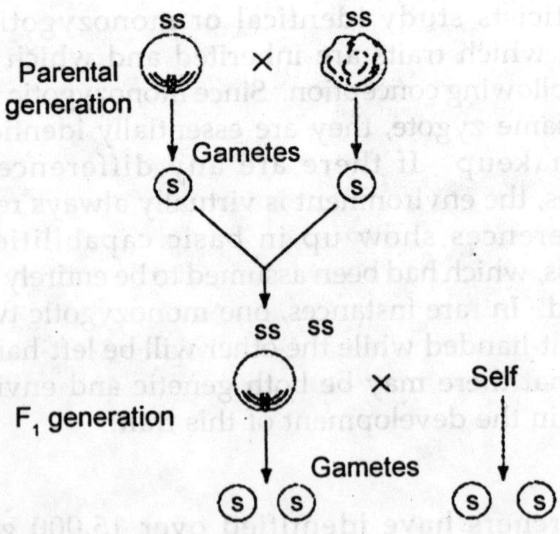

Fig. Law of Segregation.

The Law of Segregation, also known as Mendel's First Law, essentially has four parts.

1. Alternative versions of genes account for variations in inherited characteristics. This is the concept of alleles. Alleles are different versions of genes that impart the same characteristic. For example, each human has a gene that controls eye colour, but there are variations among these genes in accordance with the specific colour the gene "codes" for.
2. For each characteristic, an organism inherits two alleles, one from each parent. This means that when somatic cells are produced from two gametes, one allele comes from the mother and one from the father. These alleles may be the same (true-breeding organisms/homozygous e.g. ww and rr or different
3. If the two alleles differ, then one, the allele that encodes the dominant trait, is fully expressed in the organism's appearance; the other, the allele encoding the recessive trait, has no noticeable effect on the organism's appearance. In other words, only the dominant trait is seen in the phenotype of the organism. This allows recessive traits to be passed on to offspring even if they are not expressed. Not all traits have a dominant-recessive relationship, however. The "Japanese wonder flower" Mirabilis jalapa illustrates incomplete dominance. There is also codominance e.g. Human blood types where A and B are codominant and O is recessive.
4. The two alleles for each characteristic segregate during gamete production. This means that each gamete will contain only one allele for each gene. This allows the maternal and paternal alleles to be combined in the offspring, ensuring variation.

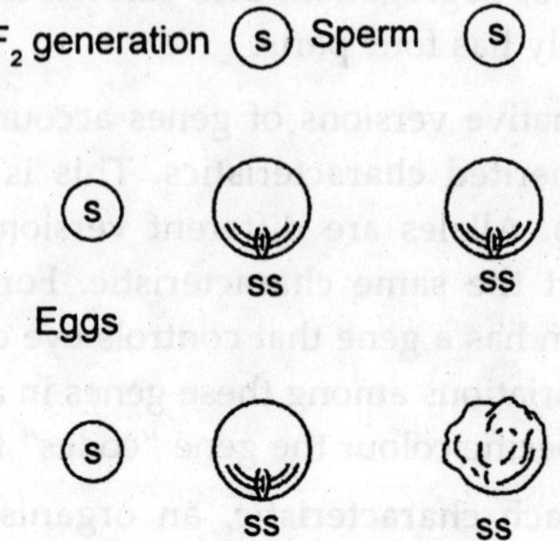

Fig.Punnett square

Punnett square explaining the behaviour of the S and s alleles.

P1: smooth X wrinkled

F1: all smooth

F2: 5474 smooth and 1850 wrinkled

Meiosis, a process unknown in Mendel's day, explains how the traits are inherited.

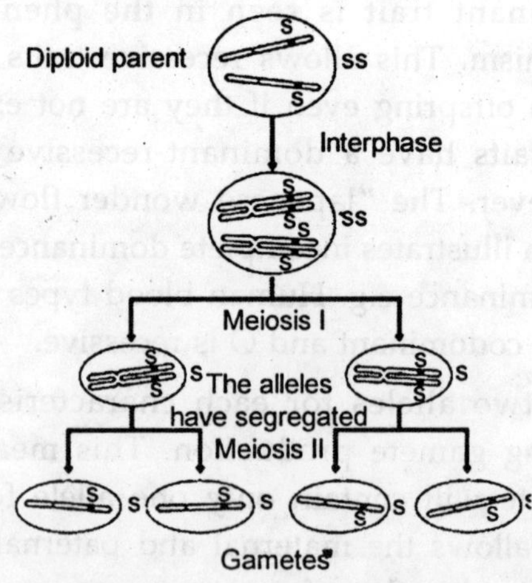

Fig.The inheritance of the S and s alleles

N.B It is often misconstrued that the gene itself is dominant, recessive, codominant, or incompletely dominant. It is, however the trait, or gene product that the allele encodes that is dominant, etc.

Law of Independent Assortment

The Law of Independent Assortment, also known as Mendel's Second Law, states that the emergence of one trait will not affect the emergence of another. While his experiments with mixing one trait always resulted in a 3:1 ratio between dominant and recessive phenotypes, his experiments with mixing two traits (dihybrid cross) showed 9:3:3:1 ratios. Mendel concluded that different traits are inherited independently of each other, so that there is no relation, for example, between a cat's colour and tail length. This is actually only true for genes that are not linked to each other.

1. Parental generation 2. F_1 generation.
3. F_2 generation. Dominant (red) and recessive (white) phenotype look alike in the F_1 (first) generation and show a 3:1 ratio in the F_2 (second) generation

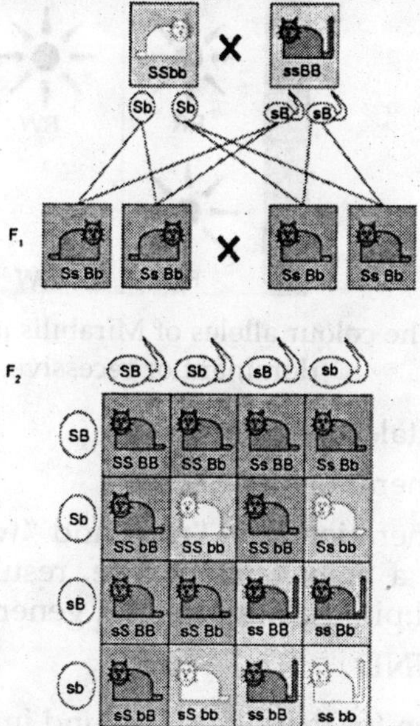

Fig: The genotypes of two independent traits show a 9:3:3:1 ratio in the F_2 generation. In this example, coat colour is indicated by B (brown, dominant) or b (white) while tail length is indicated by S (short, dominant) or s (long). When parents are homozygous for each trait ('SSbb and ssBB), their children in the F_1 generation are heterzygous at both loci and only show the dominant phenotypes.

If the children mate with each other, in the F_2 generation all combination of coat colour and tail length occur: 9 are brown/short (purple boxes), 3 are white/short (pink boxes), 3 are brown/long (blue boxes) and 1 is white/long (green box).

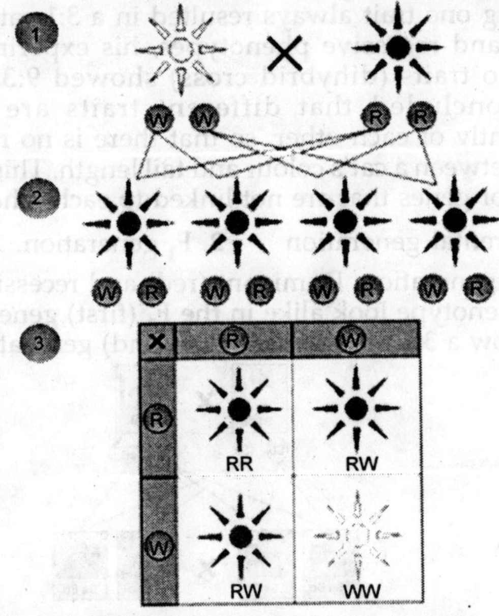

Fig: The colour alleles of Mirabilis jalapa are not dominant or recessive.

1. Parental generation.
2. F_1 generation.
3. F_2 generation. The "red" and "white" allele together make a "pink" phenotype, resulting in a 1:2:1 ratio of red:pink:white in the F_2 generation.

BACKGROUND

The reason for these laws is found in the nature of the cell nucleus. It is made up of several chromosomes carrying the genetic traits. In a normal cell, each of these chromosomes has two parts, the chromatids. A reproductive cell, which is created in a process called meiosis, usually contains only one of those chromatids of each chromosome. By merging two of these cells (usually one male and one female), the full set is restored and the genes are mixed. The resulting cell becomes a new embryo. The fact that this new life has half the genes of each parent is one reason for the Mendelian laws. The second most important reason is the varying dominance of different genes, causing some traits to appear unevenly instead of averaging out (whereby dominant doesn't mean more likely to reproduce - recessive genes can become the most common, too).

There are several advantages of this method (sexual reproduction) over reproduction without genetic exchange:

1. Instead of nearly identical copies of an organism, a broad range of offspring develops, allowing more different abilities and evolutionary strategies.
2. There are usually some errors in every cell nucleus. Copying the genes usually adds more of them. By distributing them randomly over different chromosomes and mixing the genes, such errors will be distributed unevenly over the different children. Some of them will therefore have only very few such problems. This helps reduce problems with copying errors somewhat.
3. Genes can spread faster from one part of a population to another. This is for instance useful if there's a temporary isolation of two groups. New genes developing in each of the populations don't get reduced to half when one side replaces the other, they mix and form a population with the advantages of both sides.
4. Sometimes, a mutation (e. g. Sickle Cell Anemia) can have positive side effects (in this case Malaria resistance). The mechanism behind the Mendelian laws can make it possible for some offspring to carry the advantages without the disadvantages until further mutations solve the problems.

MENDELIAN TRAIT

A Mendelian trait is one that is controlled by a single locus and shows a simple Mendelian inheritance pattern. In such cases, a mutation in a single gene can cause a disease that is inherited according to Mendel's laws. Examples include sickle-cell anemia, Tay-Sachs disease, cystic fibrosis and xeroderma pigmentosa. A disease controlled by a single gene contrasts with a multi-factorial disease, like arthritis, which is affected by several loci (and the environment) as well as those diseases inherited in a non-Mendelian fashion. The Mendelian Inheritance in Man database is a catalog of, among other things, genes in which Mendelian mutants causes disease.

Mendel studied seven traits which appeared in two discrete forms, rather than continuous characters which are often difficult to distinguish. When "truebreeding" tall plants were crossed with "true-breeding" short plants, all of the offspring were tall plants. The parents in the cross were the P1 generation, and the offspring represented the F1 generation. The trait referred to as tall was considered dominant, while short was recessive. Dominant traits were defined by Mendel as those which appeared in the F1 generation in crosses between true-breeding strains. Recessives were those which "skipped" a generation, being expressed only when the dominant trait is absent. Mendel's plants exhibited complete dominance, in which the phenotypic expression of alleles was either dominant or recessive, not "in between".

When members of the F1 generation were crossed, Mendel recovered mostly tall offspring, with some short ones also occurring. Upon statistically analyzing the F2 generation, Mendel determined the ratio of tall to short plants was approximately 3:1. Short plants have skipped the F1 generation, and show up in the F2 and succeeding generations. Mendel concluded that the traits under study were governed by discrete (separable) factors. The factors were inherited in pairs, with each generation having a pair of trait factors. We now refer to these trait factors as alleles. Having traits inherited in pairs allows for the observed phenomena of traits "skipping" generations.

Summary of Mendel's Results:

1. The F1 offspring showed only one of the two parental traits, and always the same trait.
2. Results were always the same regardless of which parent donated the pollen (was male).
3. The trait not shown in the F1 reappeared in the F2 in about 25% of the offspring.
4. Traits remained unchanged when passed to offspring: they did not blend in any offspring but behaved as separate units.
5. Reciprocal crosses showed each parent made an equal contribution to the offspring.

Mendel's Conclusions:

1. Evidence indicated factors could be hidden or unexpressed, these are the recessive traits.
2. The term phenotype refers to the outward appearance of a trait, while the term genotype is used for the genetic makeup of an organism.
3. Male and female contributed equally to the offsprings' genetic makeup: therefore the number of traits was probably two (the simplest solution).
4. Upper case letters are traditionally used to denote dominant traits, lower case letters for recessives.

Mendel reasoned that factors must segregate from each other during gamete formation (remember, meiosis was not yet known!) to retain the number of traits at 2. The Principle of Segregation proposes the separation of paired factors during gamete formation, with each gamete receiving one or the other factor, usually not both. Organisms carry two alleles for every trait. These traits separate during the formation of gametes.

Dihybrid Crosses

When Mendel considered two traits per cross (dihybrid, as opposed to singletrait-crosses, monohybrid), The resulting (F2) generation did not have 3:1 dominant:recessive phenotype ratios. The two traits, if considered to inherit independently, fit into the principle of segregation. Instead of 4 possible genotypes from a monohybrid cross, dihybrid crosses have as many as 16 possible genotypes.

Mendel realized the need to conduct his experiments on more complex situations. He performed experiments tracking two seed traits: shape and colour. A cross concerning two traits is known as a dihybrid cross.

Crosses With Two Traits

- Smooth seeds (S) are dominant over wrinkled (s) seeds.
- Yellow seed colour (Y) is dominant over green (g).

♂ gametes

gametes	RY $\frac{1}{4}$	Ry $\frac{1}{4}$	ry $\frac{1}{4}$	rY $\frac{1}{4}$
RY $\frac{1}{4}$	RR YY $\frac{1}{16}$	RR Yy $\frac{1}{16}$	Rr Yy $\frac{1}{16}$	Rr YY $\frac{1}{16}$
Ry $\frac{1}{4}$	RR Yy $\frac{1}{16}$	RR yy $\frac{1}{16}$	Rr yy $\frac{1}{16}$	Rr Yy $\frac{1}{16}$
ry $\frac{1}{4}$	Rr Yy $\frac{1}{16}$	Rr yy $\frac{1}{16}$	rr yy $\frac{1}{16}$	rr Yy $\frac{1}{16}$
rY $\frac{1}{4}$	Rr yy $\frac{1}{16}$	Rr Yy $\frac{1}{16}$	rr Yy $\frac{1}{16}$	rr yy $\frac{1}{16}$

9 : 3 : 3 : 1

Round yellow
Wrinkled, yellow
Round green
Wrinkled, green

Fig. Dihybrid Cross

Methods, Results, and Conclusions

Mendel started with true-breeding plants that had smooth, yellow seeds and crossed them with true-breeding plants having green, wrinkled seeds. All seeds in the F1 had smooth yellow seeds. The F2 plants self-fertilized, and produced four phenotypes:

315 smooth yellow
108 smooth green
101 wrinkled yellow
32 wrinkled green

Mendel analyzed each trait for separate inheritance as if the other trait were not present.The 3:1 ratio was seen separately and was in accordance with the Principle of Segregation. The segregation of S and s alleles must have happened independently of the segregation of Y and y alleles. The chance of any gamete having a Y is 1/2; the chance of any one gamete having a S is 1/2.The chance of a gamete having both Y and S is the product of their individual chances (or 1/2 X 1/2 = 1/4). The chance of two gametes forming any given genotype is 1/4 X 1/4 (remember, the product of their individual chances). Thus, the Punnett Square has 16 boxes. Since there are more possible combinations to produce a smooth yellow phenotype (SSYY, SsYy, SsYY, and SSYy), that phenotype is more common in the F2.

From the results of the second experiment, Mendel formulated the Principle of Independent Assortment — that when gametes are formed, alleles assort independently. If traits assort independent of each other during gamete formation, the results of the dihybrid cross can make sense. Since Mendel's time, scientists have discovered chromosomes and DNA. We now interpret the Principle of Independent Assortment as alleles of genes on different chromosomes are inherited independently during the formation of gametes. This was not known to Mendel.

Punnett squares deal only with probability of a genotype showing up in the next generation. Usually if enough offspring are produced, Mendelian ratios will also be produced.

Step 1 - definition of alleles and determination of dominance.

Step 2 - determination of alleles present in all different types of gametes.

Step 3 - construction of the square.

Step 4 - recombination of alleles into each small square.

Step 5 - Determination of Genotype and Phenotype ratios in the next generation.

Step 6 - Labeling of generations, for example P1, F1, etc.

While answering genetics problems, there are certain forms and protocols that will make unintelligible problems easier to do. The term "true-breeding strain" is a code word for homozygous. Dominant alleles are those that show up in the next generation in crosses between two different "true-breeding strains". The key to any genetics problem is the recessive phenotype (more properly the phenotype that represents the recessive genotype). It is that organism whose genotype can be determined by examination of the phenotype. Usually homozygous dominant and heterozygous individuals have identical phenotypes (although their genotypes are different). This becomes even more important in dihybrid crosses.

Assortment and interactions of multiple genes

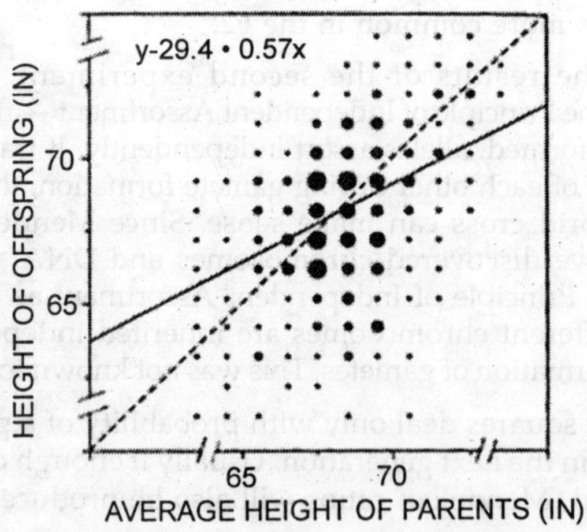

Fig.Law of independent assortment

Human height is a complex genetic trait. Francis Galton's data from 1889 shows the relationship between offsping height as a function of mean parent height. While correlated, the remaining variation in offspring heights indicates environment is also an important factor in this trait.

Organisms have thousands of genes, and in diploid organism assortment of these genes are generally independent of each other. This means that the inheritance of an allele for yellow or green pea colour is unrelated to the inheritance of alleles for white or purple flowers. This phenomenon, known as "Mendel's second law" or the "Law of independent assortment", means that the alleles of different genes get shuffled between parents to form children with many different combinations.

Often different genes can interact in a way that influences the same trait. In the blue-eyed Mary, for example, there exists a gene with alleles that determine the colour of flowers: blue or magenta. Another gene, however, controls whether the flowers have colour at all: colour or white. When a plant has two copies of this white allele, its flowers are white regardless of whether the first gene has blue or magenta alleles. This interaction between genes is called "epistasis", with the second gene epistatic to the first

Many traits are not discrete features (eg. purple or white flowers) but are instead continuous features (eg. human height and skin colour). These "complex traits" are the product of interactions of many genes. The influence of these genes is mediated, to varying degrees, by the enviroment an organism has experienced. The degree to which an organism's genes contribute to a complex trait is called "heritability". Measurement of the heritability of a trait is relative, though — in a more variable environment, the environment has a bigger influence on the total variation of the trait. For example, human height is a complex trait with a heritability of 89% in the United States. In Nigeria, however, where people experience a more variable access to good nutrition and health care, height has a heritability of only 62%.

DNA and the genetic code

The molecular basis for genes is deoxyribonucleic acid (DNA). DNA is composed of a chain of nucleotides, of which there are four types: adenine (A), cytosine (C), guanine (G), and thymine (T). Genetic information exists in the sequence of these nucleotides, and genes exist as stretches of sequence along the DNA chain. Viruses are a small exception — the similar molecule RNA instead of DNA is often the genetic material of a virus. In all non-virus organisms, which are composed of cells, each cell contains a full copy of that organism's DNA, called its genome.

In the cell, DNA exists as a double-stranded molecule, coiled into the shape of a doublehelix. Each nucleotide in DNA preferentially pairs with its partner nucleotide on the opposite strand: A pairs with T, and C pairs with G. Thus, in its two-stranded form, each strand effectively contains all necessary information, redundant with its partner strand. This structure of DNA is the physical basis for inheritance: DNA replication duplicates the genetic information by splitting the strands and using each strand as a template for a partner strand.

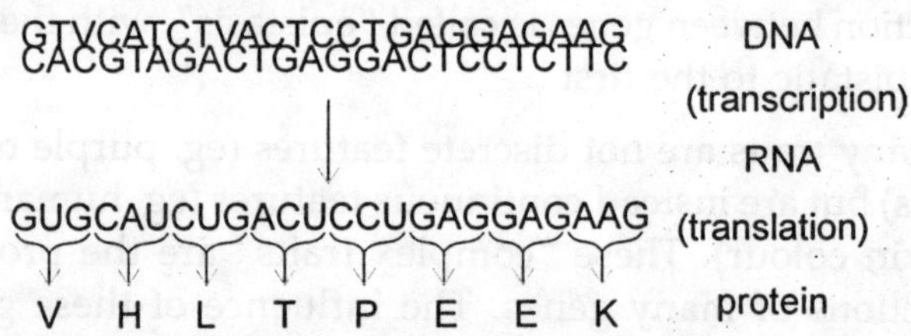

Fig. The genetic code: DNA, through a messenger RNA intermediate, codes for protein with a triplet code.

Genes express their functional effect through the production of proteins, which are complex molecules responsible for most functions in the cell. Proteins are chains of amino acids, and the DNA sequence of a gene (through an RNA intermediate) is used to produce a specific protein sequence. Each group of three nucleotides in the sequence, called a codon, corresponds to one of the twenty possibly amino acids in protein — this correspondence is called the genetic code. The specific sequence of amino acids results in a unique three-dimensional structure for that protein, thereby determining its behaviour and function.

CHROMOSOMES, RECOMBINATION, AND LINKAGE

Genes are arranged linearly along the long chains of DNA sequence, called chromosomes. In bacteria, each cell has a single circular chromosome, while eukaryotic organisms (which includes plants and animals) have their DNA arranged in multiple linear chromosomes. These DNA strands are often extremely long; the largest human chromosome, for example, is about 140 million base pairs in length.The DNA of a chromosome is associated with structural proteins which organize, compact, and control access to the DNA, forming a material called chromatin; in eukaryotes chromatin is usually composed of nucleosomes, repeating units of DNA wound around a core of histone proteins·

While haploid organisms have only one copy of each chromosome, most animals and many plants are diploid, containing two copies of each chromosome and thus two copies of every gene. An exception exists in the sex chromosomes, specialized chromosomes many animals have evolved that play a role in determining the sex of the organism. On the other chromosomes (called autosomes), the two alleles for a gene are located on identical loci of sister chromatids.

The diploid nature of chromosomes allows for genes on different chromosomes to assort independently, thereby recombining to form new combinations of genes. Genes on the same chromosome would theoretically never recombine, however, were it not for the process of chromosomal crossover. During crossover, chromosomes exchange stretches of DNA, effectively shuffling the gene alleles between the chromosomes. This process of chromosomal crossover generally occurs during meiosis, a series of cell divisions that creates haploid germ cells which later combine with other germ cells to form child organisms.

The probability of chromosomal crossover occurring between two given points on the chromosome is related to the distance between them. For an arbitrarily long distance, the probability of crossover is high enough that the inheritance of the genes is effectively uncorrelated. For genes that are closer together, however, the lower probability of crossover means that the genes demonstrate genetic linkage. The linkages between a series of genes can be combined to form a linear linkage map that roughly describes the arrangement of the genes along the chromosome.

Epigenetic inheritance

Although DNA is the genetic material of life, there are some aspects of cells that are not encoded in DNA sequence (eg. chromatin and DNA modifications) that are inherited by daughter cells when cells divide. These features are called "epigenetic" — the prefix "epi-" means "on top of" or "in addition to". Epigenetic features create differences between cells sharing the same genome, allowing cells to differentiate into different tissues in multicellular organisms. Although epigenetic features are generally dynamic over the development of organisms, some, like the phenomenon of paramutation, have multigenerational inheritance.

MODEL ORGANISMS AND GENETICS

Although geneticists originally studied inheritance in a wide range of organisms, researchers began to specialize in studying the genetics of a particular subset of organisms. The fact that significant research already existed for a given organism would encourage new researchers to choose it for further study, and so eventually a few "model organisms" became the basis for most genetics research. Common research topics in model organism genetics include the study of gene regulation and the involvement of genes in development and cancer.

Organisms were chosen, in part, for convenience — short generation times and facile genetic manipulation made some organisms popular genetics research tools. Widely used model organisms include the gut bacterium Escherichia coli, the plant Arabidopsis thaliana, baker's yeast (Saccharomyces cerevisiae), the nematode Caenorhabditis elegans, the common fruit fly (Drosophila melanogaster), and the common house mouse (Mus musculus).

MEDICAL GENETICS RESEARCH

Medical genetics research seeks to find and study the genetic causes of human diseases. When searching for an unknown gene that may be involved in a disease, researchers commonly use genetics linkage and genetic pedigree charts to find the location on the genome associated with the disease. At the population level, researchers take advantage of Mendelian randomization to look for locations in the genome that are associated with diseases, a technique especially useful for multigenic traits not clearly defined by a single gene. Once a candidate gene is found, further research is often done on the same gene (called an orthologous gene) in model organisms.

GENETIC TECHNOLOGIES

A variety of techniques exist for manipulating DNA in the laboratory. Restriction enzymes are a commonly used enzyme that cuts DNA at specific sequences, producing predictable fragments of DNA. The use of ligation enzymes allows these fragments to be stitched back together, and by ligating fragments of DNA together from different sources, researchers can create recombinant DNA. Often associated with genetically modified organisms, recombinant DNA is commonly used in the context of plasmids — short circular DNA fragments with a few genes on them. By inserting plasmids into bacteria and growing those bacteria on plates of agar (to isolate clones of bacteria cells), researchers can clonally amplify the inserted fragment of DNA (a process known as molecular cloning). (Cloning can also refer to the creation of clonal organisms, through various techniques.)

DNA can also be amplified using a procedure called the polymerase chain reaction (PCR). By using specific short sequences of DNA, PCR can exponentially amplify a targeted region of DNA. Because it can amplify from extremely small amounts of DNA, PCR is often used to detect the presence of specific DNA sequences.

DNA SEQUENCING AND GENOMICS

One of the most fundamental technologies developed to study genetics, DNA sequencing allows researchers to determine the sequence of nucleotides in DNA fragments. Developed in 1978 by Frederick Sanger and coworkers, chain-termination sequencing is now routinely used to sequence DNA fragments. With this technology, researchers have been able to study the molecular sequences associated with many human diseases. As sequencing has become less expensive and with the aid of computational tools, researchers have sequenced the genomes of many organisms by stitching together the sequences of many different fragments (a process called "genome assembly"). The Human genome project was completed in 2005, sequencing the genome to 92% coverage.

The large amount of sequences available has created the field of "genomics", research which uses computational tools to search for and analyze patterns in the full genomes of organisms. Genomics can also be considered a subfield of bioinformatics, which uses computational approaches to analyze of large sets of biological data.

DNA sequencing

The term DNA sequencing encompasses biochemical methods for determining the order of the nucleotide bases, adenine, guanine, cytosine, and thymine, in a DNA oligonucleotide. The sequence of DNA constitutes the heritable genetic information in nuclei, plasmids, mitochondria, and chloroplasts that forms the basis for the developmental programmes of all living organisms. Determining the DNA sequence is therefore useful in basic research studying fundamental biological processes, as well as in applied fields such as diagnostic or forensic research. The advent of DNA sequencing has significantly accelerated biological research and discovery. The rapid speed of sequencing attainable with modern DNA sequencing technology has been instrumental in the large-scale sequencing of the human genome, in the Human

Genome Project. Related projects, often by scientific collaboration across continents, have generated the complete DNA sequences of many animal, plant, and microbial genomes.

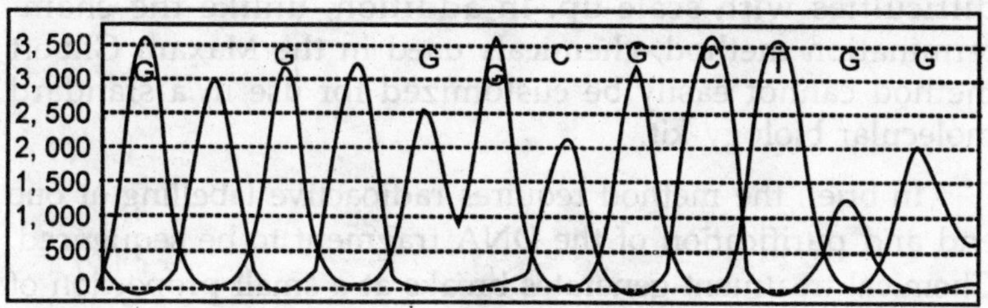

Fig.DNA Sequence Trace

Early methods

For thirty years, a large proportion of DNA sequencing has been carried out with the chain-termination method, developed by Frederick Sanger and coworkers in 1975. Prior to the development of rapid DNA sequencing methods in the early 1970s by Sanger in England and Gilbert et al. at Harvard,, a number of laborious methods were used. For instance, in 1973 Gilbert and Maxam reported the sequence of 24 basepairs using a method known as wandering-spot analysis.

It is noteworthy that RNA sequencing, which for technical reasons is easier to perform than DNA sequencing, could be considered one of the earliest forms of nucleotide sequencing. The major landmark of RNA sequencing, dating from the pre-recombinant DNA era, is the sequence of the phage MS2 genome, identified and published by Walter Fiers and coworkers.

Maxam-Gilbert sequencing

In 1976-1977, Allan Maxam and Walter Gilbert developed a DNA sequencing method based on chemical modification of DNA and subsequent cleavage at specific bases. Although Maxam and Gilbert published their chemical sequencing method two years after the ground-breaking paper of Sanger and Coulson on plus-minus sequencing, Maxam-Gilbert sequencing rapidly became more popular, since purified DNA could be used directly, while the initial Sanger method required that each read start be cloned for production of single-stranded DNA. However, with the development and improvement of the chain-termination method, Maxam-Gilbert sequencing has fallen out of favour due to its technical complexity, extensive use of hazardous chemicals, and difficulties with scale-up. In addition, unlike the chain-termination method, chemicals used in the Maxam Gilbert method cannot easily be customized for use in a standard molecular biology kit.

In brief, the method requires radioactive labelling at one end and purification of the DNA fragment to be sequenced. Chemical treatment generates breaks at a small proportion of one or two of the four nucleotide bases in each of four reactions (G, A+G, C, C+T). Thus a series of labelled fragments is generated, from the radiolabelled end to the first 'cut' site in each molecule. The fragments are then size-separated by gel electrophoresis, with the four reactions arranged side by side. To visualize the fragments generated in each reaction, the gel is exposed to X-ray film for autoradiography, yielding an image of a series of dark 'bands' corresponding to the radiolabelled DNA fragments, from which the sequence may be inferred.

Also sometimes known as 'chemical sequencing', this method originated in the study of DNA-protein interactions (footprinting), nucleic acid structure and epigenetic modifications to DNA, and within these it still has important applications.

Chain-termination Methods

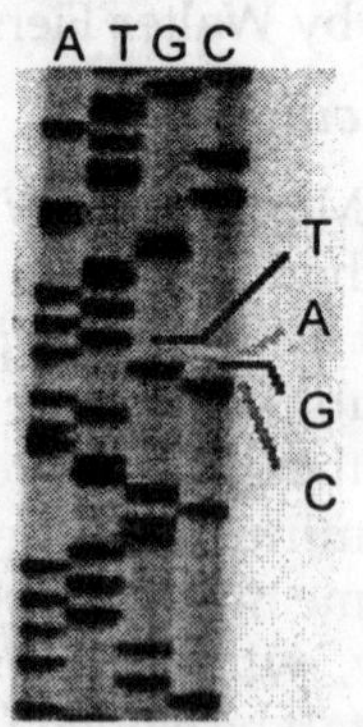

Fig.Part of a radioactively labelled sequencing gel

While the chemical sequencing method of Maxam and Gilbert, and the plus-minus method of Sanger and Coulson were orders of magnitude faster than previous methods, the chain-terminator method developed by Sanger was even more efficient, and rapidly became the method of choice. The Maxam-Gilbert technique requires the use of highly toxic chemicals, and large amounts of radiolabeled DNA, whereas the chain-terminator method uses fewer toxic chemicals and lower amounts of radioactivity. The key principle of the Sanger method was the use of dideoxynucleotides triphosphates (ddNTPs) as DNA chain terminators.

The classical chain-termination or Sanger method requires a single-stranded DNA template, a DNA primer, a DNA polymerase, radioactively or fluorescently labeled nucleotides, and modified nucleotides that terminate DNA strand elongation. The DNA sample is divided into four separate sequencing reactions, containing the four standard deoxynucleotides (dATP, dGTP, dCTP and dTTP) and the DNA polymerase. To each reaction is added only one of the four dideoxynucleotides (ddATP, ddGTP, ddCTP, or ddTTP). These dideoxynucleotides are the chain-terminating nucleotides, lacking a 3'-OH group required for the formation of a phosphodiester bond between two nucleotides during DNA strand elongation. Incorporation of a dideoxynucleotide into the nascent (elongating) DNA strand therefore terminates DNA strand extension, resulting in various DNA fragments of varying length. The dideoxynucleotides are added at lower concentration than the standard deoxynucleotides to allow strand elongation sufficient for sequence analysis.

The newly synthesized and labeled DNA fragments are heat denatured, and separated by size (with a resolution of just one nucleotide) by gel electrophoresis on a denaturing polyacrylamide-urea gel. Each of the four DNA synthesis reactions is run in one of four individual lanes (lanes A, T, G, C); the DNA bands are then visualized by autoradiography or UV light, and the DNA sequence can be directly read off the X-ray film or gel image. In the image on the right, X-ray film was exposed to the gel, and the dark bands correspond to DNA fragments of different lengths. A dark band in a lane indicates a DNA fragment that is the result of chain termination after incorporation of a dideoxynucleotide (ddATP, ddGTP, ddCTP, or ddTTP). The terminal nucleotide base can be identified according to which dideoxynucleotide was added in the reaction giving that band. The relative positions of the different bands among the four lanes are then used to read (from bottom to top) the DNA sequence as indicated.

There are some technical variations of chain-termination sequencing. In one method, the DNA fragments are tagged with nucleotides containing radioactive phosphorus for radiolabelling. Alternatively, a primer labeled at the 5′ end with a fluorescent dye is used for the tagging. Four separate reactions are still required, but DNA fragments with dye labels can be read using an optical system, facilitating faster and more economical analysis and automation. This approach is known as 'dye-primer sequencing'. The later development by L Hood and coworkers of fluorescently labeled ddNTPs and primers set the stage for automated, high-throughput DNA sequencing.

Sequence ladder by radioactive sequencing compared to fluorescent peaks (click to expand)

The different chain-termination methods have greatly simplified the amount of work and planning needed for DNA sequencing. For example, the chain-termination-based "Sequenase" kit from USB Biochemicals contains most of the reagents needed for sequencing, prealiquoted and ready to use. Some sequencing problems can occur with the Sanger Method, such as non-specific binding of the primer to the DNA, affecting accurate read out of the DNA sequence. In addition, secondary structures within the DNA template, or contaminating RNA randomly priming at the DNA template can also affect the fidelity of the obtained sequence. Other contaminants affecting the reaction may consist of extraneous DNA or inhibitors of the DNA polymerase.

Dye-terminator sequencing

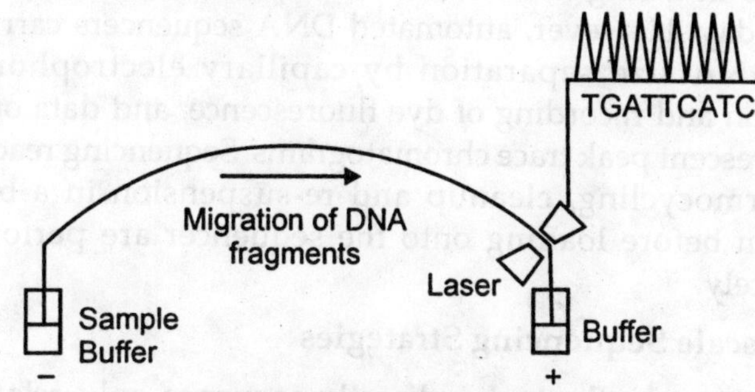

Fig. Capillary electrophoresis

An alternative to primer labelling is labelling of the chain terminators, a method commonly called 'dye-terminator sequencing'. The major advantage of this method is that the sequencing can be performed in a single reaction, rather than four reactions as in the labelled-primer method. In dye-terminator sequencing, each of the four dideoxynucleotide chain terminators is labelled with a different fluorescent dye, each fluorescing at a different wavelength. This method is attractive because of its greater expediency and speed and is now the mainstay in automated sequencing with computer controlled sequence analyzers. Its potential limitations include dye effects due to differences in the incorporation of the dye-labelled chain terminators into the DNA fragment, resulting in unequal peak heights and shapes in the electronic DNA sequence trace chromatogram after capillary electrophoresis. This problem has largely been overcome with the introduction of new DNA polymerase enzyme systems and dyes that minimize incorporation variability. The dye-terminator sequencing method, along with automated high-throughput DNA sequence analyzers, is now being used for the vast majority of sequencing projects, as it is both easier to perform and lower in cost than most previous sequencing methods.

Automation and Sample Preparation

Modern automated DNA sequencing instruments (DNA sequencers) can sequence up to 384 fluorescently labelled samples in a single batch (run) and perform as many as 24 runs a day. However, automated DNA sequencers carry out only DNA size separation by capillary electrophoresis, detection and recording of dye fluorescence, and data output as fluorescent peak trace chromatograms. Sequencing reactions by thermocycling, cleanup and re-suspension in a buffer solution before loading onto the sequencer are performed separately.

Large-scale Sequencing Strategies

Current methods can directly sequence only relatively short (300-1000 nucleotides long) DNA fragments in a single reaction. The main obstacle to sequencing DNA fragments above this size limit is insufficient power of separation for resolving large DNA fragments that differ in length by only one nucleotide. Limitations on ddNTP incorporation were largely solved by Tabor at Harvard Medical, Carl Fuller at USB biochemicals, and their coworkers.

Large-scale sequencing aims at sequencing very long DNA fragments. Even relatively small bacterial genomes contain millions of nucleotides, and the human chromosome alone contains about 246 million bases. Therefore, some approaches consist of cutting (with restriction enzymes) or shearing (with mechanical forces) large DNA fragments into shorter DNA fragments. The fragmented DNA is cloned into a DNA vector, usually a bacterial plasmid, and amplified in Escherichia coli. The amplified DNA can then be purified from the bacterial cells (a disadvantage of bacterial clones for sequencing is that some DNA sequences may be inherently un-clonable in some or all available bacterial strains, due to deleterious effect of the cloned sequence on the host bacterium or other effects). These short DNA fragments purified from individual bacterial colonies are then individually and completely sequenced and assembled electronically into one long, contiguous sequence by identifying 100%-identical overlapping sequences between them (shotgun sequencing). This method does not require any pre-existing information about the sequence of the DNA and is often referred to as de novo sequencing. Gaps in the assembled sequence may be filled by Primer walking, often with sub-cloning steps (or transposon-based sequencing depending on the size of the remaining region to be sequenced). These strategies all involve taking many small reads of the DNA by one of the above methods and subsequently assembling them into a contiguous sequence. The different strategies have different tradeoffs in speed and accuracy; the shotgun method is the most practical for sequencing large genomes, but its assembly process is complex and potentially error-prone - particularly in the presence of sequence repeats.

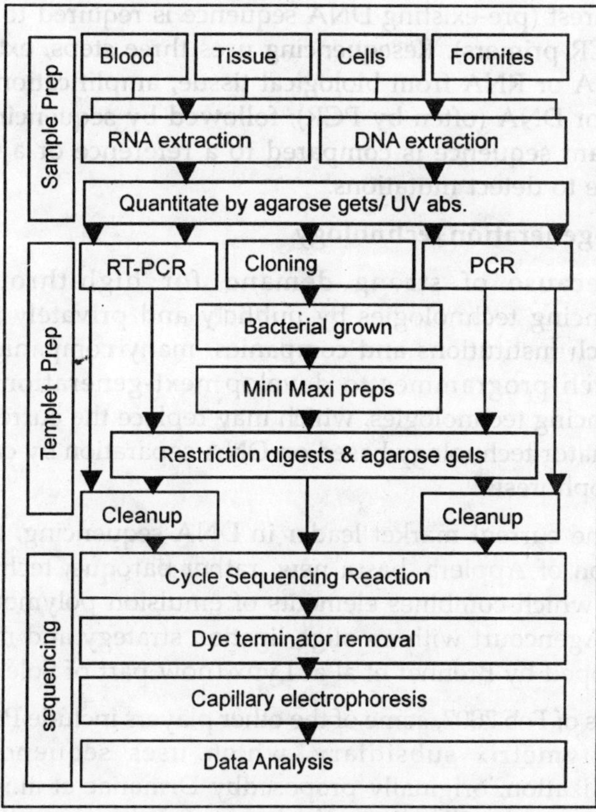

Fig. Resequencing steps. Sample prep: Extraction of nucleic acid. Template prep: Amplification and preparation of a small region of the target region.

The human genome is about 3 billion (3,000,000,000) bp long; if the average fragment length is 500 bases, it would take a minimum of six million (3 billion/500) to sequence the human genome (not allowing for overlap = 1-fold coverage). Keeping track of such a high number of sequences presents significant challenges, only held down by developing and coordinating several procedural and computational algorithms, such as efficient database development and management.

Resequencing or targeted sequencing is utilized for determining a change in DNA sequence from a "reference" sequence. It is often performed using PCR to amplify the region of interest (pre-existing DNA sequence is required to design the PCR primers). Resequencing uses three steps, extraction of DNA or RNA from biological tissue; amplification of the RNA or DNA (often by PCR); followed by sequencing. The resultant sequence is compared to a reference or a normal sample to detect mutations.

Next-generation technology

Because of strong demand for high-throughput sequencing technologies by publicly and privately funded research institutions and companies, many companies have research programmes to develop next-generation DNA-sequencing technologies, which may replace the current dye-terminator technology based on DNA separation by capillary electrophoresis.

The current market leader in DNA sequencing, the ABI division of Applera, has a new, rather baroque, technology, Solid, which combines elements of emulsion polymerization from Agencourt with an oligo ligation strategy and methods developed by Brenner et al at Lynx (now part of Solexa)

As of Feb 2007, some of the other players include Perlegen, an Affymetrix subsidiary, which uses sequencing by hybridization, originally proposedby Drmanac et al.; Solexa, now owned by Illumina, which uses a Bridge amplification technology originally developed by Adams and Kron; Helicos, a private company in Cambridge, MA which emphasizes single molecule sequencing; 454 Life Sciences, a subsidiary of Curagen with funding from Roche, using the innovative pyrosequencing method developed by Ronaghi et al; a colony-based technique (from Mitra and Church at Harvard) developed by Agencourt and licensed to ABI, and probably others.

Other proposals include labeling the DNA polymerase, reading the sequence as a strand transits through nanopores, microscopy-based techniques capable of single molecule resolution, such as AFM or electron microscopy and many others. There is an extensive list of methods that have failed to become commercial successes, such as the blot-based genome sequencing method of Church and Gilbert, commercialized by Integrated Genetics, methods using tin isotope-tagged nucleotides and mass spec detectors; and the "bottom wiper" of Pohl commercialized by Hoefer.

LIMITATIONS OF CURRENT TECHNOLOGY

Claims that the complete sequence of the human genome is available are somewhat inaccurate. For example, the Lawrence Berkeley Laboratory published a paper entitled "The complete sequence of human chromosome 5". Careful inspection of this paper shows that despite the use of the word "complete", there are non-sequenced genomic regions, and that the authors can only estimate how much more work remains to be done. While some DNA regions, particularly those that are rich in simple sequence repeats such as telomeres, centromeres, and AT-rich intergenic regions are difficult or impossible to sequence with current technology, it is not clear how much of the unsequenced region on chromosome 5 resides in repeat-rich regions.

Researchers are beginning to use knowledge learned from genome sequencing research to Fig out how being healthy and being sick are different at the level of molecules. And doctors are starting to use genetic information to make treatment choices.

For example, a diagnostic test can search for differences in the level of expression of a particular gene in breast cancer cells and predict whether a person will respond to a drug called Herceptin.

The cancerous cells of some people who have breast cancer make an abundance of "HER2" proteins that are targeted by Herceptin. For those people, Herceptin is a miracle drug because it reduces the risk that their breast cancer will come back, and it also decreases their odds of dying from the disease.

For cancer patients whose tumor genes do not express HER2, Herceptin won't do a thing, though, so it shouldn't be prescribed. Research is proceeding quickly to develop other genetic tests that may help diagnose and treat a wide range of health problems beyond cancer.

INDIVIDUALIZED PRESCRIPTIONS

One way variations in our genes make a difference in our health is by affecting how our bodies react to medicines. The unsettling truth is that medicines work as expected in fewer than half of the people who take them.

While environmental and lifestyle factors can explain some of this, a good part of the individual variability in response to medicines can be attributed to variants in the genes that make cytochrome P450 proteins. These proteins process many of the drugs we take.

Because each person's set of genes is a little different, the proteins that the genes encode are also slightly different. These changes can affect how the cytochrome P450 proteins (and many other types of proteins) work on drugs.

Doctors first realized this in the 1950s, when some patients had bad—sometimes fatal—reactions to an anesthetic medicine used in surgery. Experiments revealed that those who reacted poorly had a genetic variation in the enzyme that breaks down and disposes of the anesthetic after it's been in the body for a while.

People whose genes encode the variant enzyme had no trouble at all until they needed surgery that required general anesthesia. In the operating room, a normal human genetic variation suddenly led to a medical crisis!

Fortunately, this type of serious reaction to an anesthetic is very rare. But many reactions to medicines aren't so unusual. Researchers know that genetic variations can cause some common medicines to have dangerous side effects. For example, some people who take the colon cancer drug Camptosar® (also known as irinotecan) can develop diarrhea and a life-threatening infection if they have a variant form of the gene for the protein that metabolizes Camptosar.

Genetic variations can also cause drugs to have little effect at all. For example, in some people, pain medicines containing codeine, like Tylenol® with Codeine Elixir, offer no relief because their bodies break it down in an unusual way.

The use of genetic information to predict how people will respond to medicines is called pharmacogenetics. The ultimate goal of this field of study is to customise treatments based on an individual's genes.

With this kind of approach, every patient won't be treated the same, because doctors will have the molecular tools to know ahead of time which drug, and how much of it, to prescribe—or whether to prescribe it at all.

THE HEALING POWER OF DNA

Pharmacogenetics is advancing quickly since scientists have a lot of new information from the Human Genome Project and new computer tools that help them analyze the information. One disease for which progress has been rapid is cancer.

Consider the fact that cancer is often treated with a chemotherapy "cocktail," a combination of several different medicines. Each of the drugs in the mixture interacts with different proteins that control how well that particular drug works and how quickly it is metabolized in the body. What's more, each drug may have its own set of unpleasant—even potentially life-threatening—side effects.

For these reasons, individually targeted, gene-based prescriptions for chemotherapy may offer a real benefit to people with cancer.

Currently, chemotherapy cures about 80 percent of the children who have been diagnosed with acute lymphoblastic leukemia, the most common childhood cancer. The remaining 20 percent are at risk of the cancer coming back.

Mary Relling, a research clinical pharmacist at St. Jude Children's Research Hospital in Memphis, Tennessee, discovered that variations in two genes can predict which patients with acute lymphoblastic leukemia are likely to be cured by chemotherapy. Her research team also identified more than 100 genes expressed only in cancer cells that can be used to predict resistance to chemotherapy drugs.

By taking patient and cancer cell genetic pro- files into account, Relling says, researchers can develop more effective treatments for the disease. Other pharmacogenetic scientists are studying the effects of gene variants on patients' responses to drugs used to treat AIDS, allergies, infections, asthma, heart disorders, and many other conditions.

For example, researchers recently identified two different genetic variants that play a central role in determining the body's response to Coumadin® (also known as warfarin), a widely prescribed medicine given to people who are at risk for blood clots or heart attacks. Although 2 million Americans take this blood-thinning drug every day, it is very difficult to administer, since its effects vary widely in different people who are taking the same dose. Giving the right dose is essential, because too much Coumadin can cause excessive bleeding, while too little can allow blood clots to form.

Allan Rettie, a medicinal chemist at the University of Washington in Seattle, discovered that genetic variation among people influences the activity of a protein in the blood that is Coumadin's molecular target. He and other scientists are now trying to translate these findings into a genetic test that could help doctors predict what dose of Coumadin is appropriate based on each patient's DNA profile.

Genes in human behaviour

Did you know that, in addition to traits you can see like hair colour and physique, genes also contribute to how we behave? It may come as a surprise that many researchers are answering basic questions about the genetics of behaviour by studying insects.

For example, Gene Robinson, an entomologist at the University of Illinois at Urbana Champaign, works with honeybees. Robinson says that if you look at honeybees in their natural hive environment, you'll quickly see that they are very outgoing. In fact, according to Robinson, honeybees can't survive without the social structure of their community within the hive.

This characteristic makes them a perfect species in which to study the genetics of behaviour.

What's particularly interesting about bees is that rather than being stuck in a particular job, they change jobs according to the hive's needs. Robinson has identified certain genes whose activity changes during a job shift, suggesting that the insects' environment helps to shape their gene expression.

Researchers who are beginning to understand these connections are working in a brandnew field of investigation named by Robinson himself: sociogenomics.

What does all of this mean for humans, you wonder? It underscores the fact that, far from being set in stone, our genomes are influenced by both heredity and environment, finetuned and sculpted by our social life and the things we do every day.

Cause and Effect

What more do we need to know about how genes shape who we are and what we become?

Context plays an enormous role in determining how organisms grow and develop, and what diseases they get. A unique combination of genetic and environmental factors, which interact in a way that is very hard to predict, determines what each person is like.

Very few, if any, scientists would argue with this. Whether a gene is expressed, and even whether the mRNA transcript gets translated into a protein, depends on the environment. Few diseases—most of which are very rare—are caused completely by a mutated gene.

In most cases, getting or avoiding a disease depends not just on genes but on things within your control, such as diet, exercise, and whether or not you smoke.

It will be many years before scientists clearly understand the detailed meaning of our DNA language and how it interacts with the environment in which we live. Still, it's a great idea to find out as much as you can about your family's health history. Did any of your relatives have diabetes? Do people in your family tree have cancer or heart disease?

Keep in mind that diseases such as these are relatively common, so it's pretty likely that at least one relative will have one of them. But if heart disease, diabetes, or particular types of cancer "run in your family," especially if a lot of your relatives get the condition when they are fairly young, you may want to talk with your doctor about your own risk for developing the disease.

If you do discover that you are at higher-than usual risk for a disease like breast cancer or heart disease, you may be able to prevent the disease, or delay its onset, by altering your diet, exercising more, or making other lifestyle changes. You may also be able to take advantage of screening tests like mammograms (breast X rays that detect signs of cancer), colonoscopies (imaging tests for colon cancer), or blood sugar tests for diabetes. Screening tests can catch diseases early, when treatment is most successful.

Many scientists focus on human genes, most of which have counterparts in the genomes of model organisms. However, in the case of infections caused by microorganisms, understanding how the genomes of bacteria, viruses, and parasites differ from ours is a very important area of health research.

Most of the medicines we take to treat infections by bacteria and viruses have come from scientists' search for molecular weak points in these tiny organisms. As mentioned earlier for example, some antibiotics kill bacteria by disarming their protein-making ribosomes.

So why don't they kill human cells, too? The answer is that human and bacterial ribosomes are different. Genome sequencing is a powerful tool for identifying differences that might be promising targets for new drugs.

Comparing genetic sequences in organisms that are resistant and non-resistant to drugs can reveal new approaches to fighting resistance. Drug resistance is a worldwide problem for a number of diseases, including malaria.

Although researchers have developed several different types of medicines to treat this disease—caused by parasites carried by mosquitoes, not by a bacterium or a virus—malaria is rampant, especially in the developing world.

This is partly because not all people have access to treatment, or to simple preventive measures like bed nets, which protect sleeping people from mosquito bites. But another problem is the malaria parasite itself, which has rapidly evolved ways to avoid the effects of antimalarial drugs.

Scientists are trying to counter this process by studying microbial genetic information. In the case of malaria, geneticists like Dyann Wirth of the Harvard School of Public Health compare the genomes of drug-resistant parasites and those that can still be killed by antimalarial medicines.

Wirth's research suggests that it should be possible to develop a simple, inexpensive genetic test that could be given to people with malaria, anywhere in the world. This test would identify drugs that are likely to be most effective and help decrease the rate at which parasites become resistant to the antimalarial medicines we already have.

These gangs, known as biofilms, are layers of slime that develop naturally when bacteria congregate on surfaces like stone, metal, and wood. Or on your teeth: yuck!

Biofilms grow in all sorts of conditions. For example, one biofilm known as "desert varnish" thrives on rocks, canyon walls, or, sometimes, entire mountain ranges, leaving a reddish or other-colored stain. It is thought that petroglyphs left on boulders and cave walls by early desert dwellers were often formed by scraping through the coating of desert varnish formations with a hard object.

Sometimes, biofilms perform helpful functions. One of the best examples of the use of biofilms to solve an important problem is in the cleaning of wastewater.

But biofilms can be quite harmful, contributing to a wide range of serious health problems including cholera, tuberculosis, cystic fibrosis, and food poisoning. They also underlie many conditions that are not life-threatening but are nonetheless troublesome, like tooth decay and ear infections.

Bacteria form biofilms as a survival measure. By living in big groups rather than in isolation, the organisms are able to share nutrients and conserve energy. How do they do it?

A biofilm is not just a loose clump of cells— it's a highly sophisticated structure. As in any community, the individuals in biofilms communicate with each other.

Beyond that, many aspects of biofilms are poorly understood. Bacterial geneticist Bonnie Bassler of Princeton University in New Jersey is working to understand biofilms better, with the goal of being able to use this knowledge to break up bacterial "gang meetings."

Bassler's research subjects have a definite visual appeal. They glow in the dark, but only when they are part of a group. The bioluminescence, as the glow is called, arises from chemical reactions taking place within the biofilm. It provides a way for the bacteria to talk to each other, estimate the population size of their community, and distinguish themselves from other types of microorganisms.

Through her studies, Bassler has identified a set of molecules that biofilm-forming microorganisms use to pass messages to each other. By devising genetically based methods to cut off the chatter, Bassler reasons, she may be able to cause bacterial communities to fall apart. This approach would provide a whole new way to treat health problems linked to harmful biofilms.

Chapter 3

Molecular Genetics

GENOMES

In biology the genome of an organism is the whole hereditary information of an organism that is encoded in the DNA (or, for some viruses, RNA). This includes both the genes and the non-coding sequences. The term was coined in 1920 by Hans Winkler, Professor of Botany at the University of Hamburg, Germany, as a portmanteau of the words gene and chromosome.

More precisely, the genome of an organism is a complete DNA sequence of one set of chromosomes; for example, one of the two sets that a diploid individual carries in every somatic cell. The term genome can be applied specifically to mean the complete set of nuclear DNA (i.e., the nuclear genome) but can also be applied to organelles that contain their own DNA, as with the mitochondrial genome or the chloroplast genome. When people say that the genome of a sexually reproducing species has been "sequenced," typically they are referring to a determination of the sequences of one set of autosomes and one of each type of sex chromosome, which together represent both of the possible sexes. Even in species that exist in only one sex, what is described as "a genome sequence" may be a composite from the chromosomes of various individuals. In general use, the phrase genetic makeup is sometimes used conversationally to mean the genome of a particular individual or organism. The study of the global properties of genomes of related organisms is usually referred to as genomics, which distinguishes it from genetics which generally studies the properties of single genes or groups of genes.

GENOME PROJECTS

Genome projects are scientific endeavours that ultimately aim to determine the complete genome sequence of an organism (be it an animal, a plant, a fungus, a bacterium, an archaean, a protist or a virus). The genome sequence for any organism requires the DNA sequences for each of the chromosomes in an organism to be determined. For bacteria, which usually have just one chromosome, a genome project will aim to map the sequence of that chromosome. Humans, with 22 pairs of chromosomes and 2 sex chromosomes, will require 24 separate chromosome sequences in order to represent the completed genome.

The Human Genome Project was a landmark genome project and some have argued that the era of genomics is one of the more fundamental advances in human history.

Genome Assembly

Genome assembly refers to the process of taking a large number of short DNA sequences, all of which were generated by a shotgun sequencing project, and putting them back together to create a representation of the original chromosomes from which the DNA originated. In a shotgun sequencing project, all the DNA from a source (usually a single organism, anything from a bacterium to a mammal) is first fractured into millions of small pieces. These pieces are then "read" by automated sequencing machines, which can read up to 900 nucleotides or bases at a time. (The four bases are adenine, guanine, cytosine, and thymine, represented as AGCT.) A genome assembly algorithm works by taking all the pieces and aligning them to one another, and detecting all places where two of the short sequences, or reads, overlap. These overlapping reads can be merged together, and the process continues.

Genome assembly is a very difficult computational problem, made more difficult because genomes contain large numbers of identical sequences, known as repeats. These repeats can be thousands of nucleotides long, and some occur in thousands of different locations, especially in the large genomes of plants and animals.

Assembly software

Most research institutes that sequence DNA, use their own software for assembling the sequences that they produce. Some well known assembly programmes include:

AMOS (A Modular, Open-Source assembler) is a well-known open source effort to bring together the efforts of leading genome assembly code developers. The home of AMOS is currently http://amos.sourceforge.net. AMOS was initiated at The Institute for Genomic Research by Steven Salzberg, Mihai Pop, and Art Delcher.

The Celera Assembler was the assembler developed by Gene Myers, Granger Sutton, Art Delcher, and others at Celera Genomics from 1998 until approximately 2002.

The Arachne assembler began in 2000 as the doctoral thesis of Serafim Batzoglou, now at Stanford University. Since that time, it has been developed by a team lead by David B. Jaffe at the Broad Institute, formerly part of the Whitehead Institute.

Genome annotation

Genome annotation is the process of attaching biological information to sequences. It consists of two main steps:

1. Identifying elements on the genome, a process called Gene Finding,
2. Attaching biological information to these elements.

Automatic annotation tools try to perform all this by computer analysis, as opposed to manual annotation which involves human expertise. Ideally, these approaches co-exist and complement each other in the same annotation pipeline.

The basic level of annotation is using BLAST for finding similarities, and then annotating genomes based on that. However, nowadays more and more additional information is added to the annotation platform. The additional information allows manual annotators to deconvolute discrepancies between genes that are given the same annotation.

For example, the seed database uses genome context information, similarity scores, experimental data, and integrations of other resources to provide the most accurate genome annotations through their Subsystems approach.

Structural annotation consists in the identification of genomic elements.

- ORFs and their localisation
- gene structure
- coding regions
- location of regulatory motifs

Functional annotation consists in attaching biological information to genomic elements.

- biochemical function
- biological function
- involved regulation and interactions
- expression

These steps may involve both biological experiments and in silico analysis.

A variety of software tools have been developed to permit scientists to view and share genome annotations.

Completion of Genome Projects

When sequencing a genome, there are usually regions that are difficult to sequence (often regions with highly repetitive DNA). Thus, 'completed' genome sequences are rarely ever complete, and terms such as 'working draft' or 'essentially complete' have been used to more accurately describe the status of such genome projects. Even when every base pair of a genome sequence has been determined, there are still likely to be errors present because DNA sequencing is not a completely accurate process. It could also be argued that a complete genome project should include the sequences of mitochondria and (for plants) chloroplasts as these organelles have their own genomes.

It is often reported that the goal of sequencing a genome is to obtain information about the complete set of genes in that particular genome sequence. The proportion of a genome that encodes for genes may be very small (particularly in eukaryotes such as humans, where coding DNA may only account for a few percent of the entire sequence). However, it is not always possible (or desirable) to only sequence the coding regions separately. Also, as scientists understand more about the role of this noncoding DNA (often referred to as junk DNA), it will become more important to have a complete genome sequence as a background to understanding the genetics and biology of any given organism.

In many ways genome projects do not confine themselves to only determining a DNA sequence of an organism. Such projects may also include gene prediction to find out where the genes are in a genome, and what those genes do. There may also be related projects to sequence ESTs or mRNAs to help find out where the genes actually are.

Historical and technological perspectives

Historically, when sequencing eukaryotic genomes (such as the worm Caenorhabditis elegans) it was common to first map the genome to provide a series of landmarks across the genome. Rather than sequence a chromosome in one go, it would be sequenced piece by piece (with the prior knowledge of approximately where that piece is located on the larger chromosome). Changes in technology and in particular improvements to the processing power of computers, means that genomes can now be 'shotgun sequenced' in one go (there are caveats to this approach though when compared to the traditional approach).

Improvements in DNA sequencing technology has meant that the cost of sequencing a new genome sequence has steadily fallen (in terms of cost per base pair) and newer technology has also meant that genomes can be sequenced far more quickly. When research agencies decide what new genomes to sequence, the emphasis has been on species which have either a relevance to human health (e.g. pathogenic bacteria or vectors of disease such as mosquitos) or species which have commercial importance (e.g. livestock and crop plants). Secondary emphasis is placed on species whose genomes will help answer important questions in molecular evolution (e.g. the common chimpanzee).

In the future, it is likely that it will become even cheaper and quicker to sequence a genome. This will allow for complete genome sequences to be determined from many different individuals of the same species. For humans, this will allow us to better understand aspects of human genetic diversity.

MUTATIONS AND EVOLUTION

Every organism has a set of genes and half of the genes of that organism come from each parent. The combinations of the genes causes the variation of individuals within the species. The genes of a butterfly, an ape or a fowl carry the code that determines the appearance and the character of the butterfly, the ape and the fowl. The genetic code allows an overwhelming variety within the species of the kind. A mutation is basically a gene that has an abnormality in relation to its normal configuration. The abnormality can then be passed to successive offspring, thereby producing a marked difference. There are many different types of mutations. The smallest possible genetic mutation is a 'point-mutation'. This occurs in the DNA when the base-pairs combine with the 'wrong' partner. Multiple point-mutations are common and are found to increase substantially by the effect of mutagens. Mutations are, of course, heritable and these can extend to whole or part chromosomal mutations. Because many genes are affected by a chromosomal mutation, these often have drastic ramifications on the offspring.

Mutation

In biology, mutations are changes to the base pair sequence of genetic material (either DNA or RNA). Mutations can be caused by copying errors in the genetic material during cell division and by exposure to ultraviolet or ionizing radiation, chemical mutagens, or viruses, or can occur deliberately under cellular control during processes such as meiosis or hypermutation. In multicellular organisms, mutations can be subdivided into germline mutations, which can be passed on to descendants, and somatic mutations. The somatic mutations cannot be transmitted to descendants in animals. Plants sometimes can transmit somatic mutations to their descendants asexually or sexually (in case when flower buds develop in somatically mutated part of plant).

Mutations create variations in the gene pool, and the less favorable (or deleterious) mutations are removed from the gene pool by natural selection, while more favorable (beneficial or advantageous) ones tend to accumulate, resulting in evolutionary change. For example, a butterfly may develop offspring with a new mutation caused say by ultraviolet light from the sun. In most cases, this mutation is not good, since obviously there was no 'purpose' for such change at the molecular level. However, sometimes a mutation may change, say, the butterfly's colour, making it harder for predators to see it; this is an advantage and the chances of this butterfly surviving and producing its own offspring are a little better, and over time the number of butterflies with this mutation may form a large percentage of the species. Neutral mutations are defined as mutations whose effects do not influence the fitness of either the species or the individuals who make up the species. These can accumulate over time due to genetic drift. The overwhelming majority of mutations have no significant effect, since DNA repair is able to mend most changes before they become permanent mutations, and many organisms have mechanisms for eliminating otherwise permanently mutated somatic cells.

Classification

By effect on structure

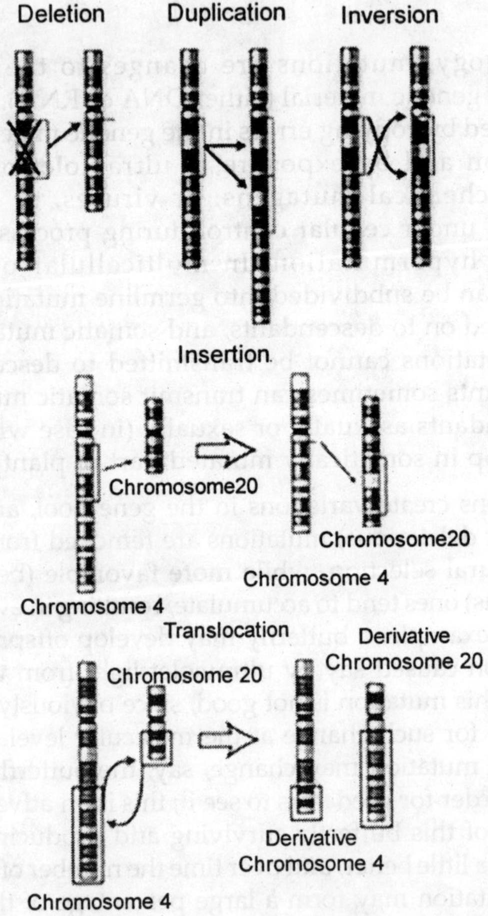

Fig. five types of chromosomal mutations.

The sequence of a gene can be altered in a number of ways. Gene mutations have varying effects on health depending on where they occur and whether they alter the function of essential proteins. Structurally, mutations can be classified as:

1. Small-scale mutations, such as affecting a small gene is one or a few nucleotides, including:

   a Point mutations, often caused by chemicals or malfunction of DNA replication, exchange a single nucleotide for another. Most common is the transition that exchanges a purine for a purine (A "! G) or a pyrimidine for a pyrimidine, (C "! T). A transition can be caused by nitrous acid, base mispairing, or mutagenic base analogs such as 5-bromo-2-deoxyuridine (BrdU). Less common is a transversion, which exchanges a purine for a pyrimidine or a pyrimidine for a purine (C/T "! A/G). A point mutation can be reversed by another point mutation, in which the nucleotide is changed back to its original state (true reversion) or by second-site reversion (a complementary mutation elsewhere that results in regained gene functionality). These changes are classified as transitions or transversions. An example of a transversion is adenine (A) being converted into a cytosine (C). There are also many other examples that can be found. Point mutations that occur within the protein coding region of a gene may be classified into three kinds, depending upon what the erroneous codon codes for:

      i Silent mutations: which code for the same amino acid.

      ii Missense mutations: which code for a different amino acid.

      iii Nonsense mutations: which code for a stop and can truncate the protein.

   b. Insertions add one or more extra nucleotides into the DNA. They are usually caused by transposable elements, or errors during replication of repeating elements (e.g. AT repeats). Insertions in the coding region of a gene may alter splicing of the mRNA (splice site mutation), or cause a shift in the reading frame (frameshift), both of which can significantly alter the gene product. Insertions can be reverted by excision of the transposable element.

c. Deletions remove one or more nucleotides from the DNA. Like insertions, these mutations can alter the reading frame of the gene. They are irreversible.

2. Large-scale mutations in chromosomal structure, including:

a Amplifications (or gene duplications) leading to multiple copies of all chromosomal regions, increasing the dosage of the genes located within them.

b Deletions of large chromosomal regions, leading to loss of the genes within those regions.

c Mutations whose effect is to juxtapose previously separate pieces of DNA, potentially bringing together separate genes to form functionally distinct fusion genes (e.g. bcr-abl). These include:

i Chromosomal translocations: interchange of genetic parts from nonhomologous chromosomes.

ii Interstitial deletions: an intra-chromosomal deletion that removes a segment of DNA from a single chromosome, thereby apposing previously distant genes. For example, cells isolated from a human astrocytoma, a type of brain tumor, were found to have a chromosomal deletion removing sequences between the "fused in glioblastoma" (fig) gene and the receptor tyrosine kinase "ros", producing a fusion protein (FIG-ROS). The abnormal FIG-ROS fusion protein has constitutively active kinase activity that causes oncogenic transformation (a transformation from normal cells to cancer cells).

iii Chromosomal inversions: reversing the orientation of a chromosomal segment.

d Loss of heterozygosity: loss of one allele, either by a deletion or recombination event, in an organism that previously had two different alleles.

By effect on function

- Loss-of-function mutations are the result of gene product having less or no function. When the allele has a complete loss of function (null allele) it is often called an amorphic mutation. Phenotypes associated with such mutations are most often recessive. Exceptions are when the organism is haploid, or when the reduced dosage of a normal gene product is not enough for a normal phenotype (this is called haploinsufficiency).
- Gain-of-function mutations change the gene product such that it gains a new and abnormal function. These mutations usually have dominant phenotypes. Often called a neomorphic mutation.
- Dominant negative mutations (also called antimorphic mutations) have an altered gene product that acts antagonistically to the wild-type allele. These mutations usually result in an altered molecular function (often inactive) and are characterised by a dominant or semi-dominant phenotype. In humans, Marfan syndrome is an example of a dominant negative mutation occurring in an autosomal dominant disease. In this condition, the defective glycoprotein product of the fibrillin gene (FBN1) antagonizes the product of the normal allele.
- Lethal mutations are mutations that lead to a phenotype incapable of effective reproduction.

By aspect of phenotype affected

- Morphological mutations usually affect the outward appearance of an individual. Mutations can change the height of a plant or change it from smooth to rough seeds.
- Biochemical mutations result in lesions stopping the enzymatic pathway. Often, morphological mutants are the direct result of a mutation due to the enzymatic pathway.

Special classes

- Conditional mutation is a mutation that has wild-type (or less severe) phenotype under certain "permissive" environmental conditions and a mutant phenotype under certain "restrictive" conditions. For example, a temperature-sensitive mutation can cause cell death at high temperature (restrictive condition), but might have no deleterious consequences at a lower temperature (permissive condition).

Causes of mutation

Two classes of mutations are spontaneous mutations (molecular decay) and induced mutations caused by mutagens.

Spontaneous mutations on the molecular level include:

- Tautomerism - A base is changed by the repositioning of a hydrogen atom.
- Depurination - Loss of a purine base (A or G).
- Deamination - Changes a normal base to an atypical base; C '! U, (which can be corrected by DNA repair mechanisms), or spontaneous deamination of 5-methycytosine (irreparable), or A '! HX (hypoxanthine).
- Transition - A purine changes to another purine, or a pyrimidine to a pyrimidine.
- Transversion - A purine becomes a pyrimidine, or vice versa.

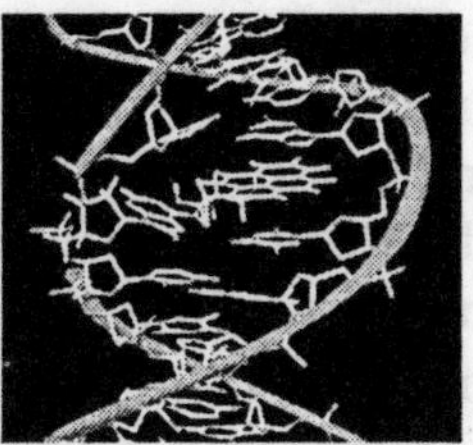

Fig. Benzopyrene, the major mutagen in tobacco smoke, in an adduct to DNA.

Induced mutations on the molecular level can be caused by:

1. Chemicals
   a. Nitrosoguanidine (NTG)
   b. Hydroxyamine NH3OH
   c. Base analogs (e.g. BrdU)
   d. Simple chemicals (e.g. acids)
   e. Alkylating agents (e.g. N-ethyl-N-nitrosourea (ENU)) These agents can mutate both replicating and non-replicating DNA. In contrast, a base analog can only mutate the DNA when the analog is incorporated in replicating the DNA. Each of these classes of chemical mutagens has certain effects that then lead to transitions, transversions, or deletions.
   f. Methylating agents (e.g. ethyl methanesulfonate (EMS))
   g. Polycyclic hydrocarbons (e.g. benzopyrenes found in internal
   h. DNA intercalating agents (e.g. ethidium bromide)
   i DNA crosslinker (e.g. platinum)
   j Oxidative damage caused by oxygen(O)] radicals
1. Radiation
   a Ultraviolet radiation (nonionizing radiation) - excites electrons to a higher energy level. DNA absorbs one form, ultraviolet light. Two

nucleotide bases in DNA - cytosine and thymine- are most vulnerable to excitation that can change base-pairing properties. UV light can induce adjacent thymine bases in a DNA strand to pair with each other, as a bulky dimer.

b Ionizing radiation

DNA has so-called hotspots, where mutations occur up to 100 times more frequently than the normal mutation rate. A hotspot can be at an unusual base, e.g., 5-methylcytosine.

Mutation rates also vary across species. Evolutionary biologists have theorized that higher mutation rates are beneficial in some situations, because they allow organisms to evolve and therefore adapt more quickly to their environments. For example, repeated exposure of bacteria to antibiotics, and selection of resistant mutants, can result in the selection of bacteria that have a much higher mutation rate than the original population (mutator strains).

Nomenclature

Nomenclature of mutations specify the type of mutation and base or amino acid changes.

- Amino acid substitution - (e.g. D111E) The first letter is the one letter code of the wildtype amino acid, the number is the position of the amino acid from the N terminus and the second letter is the one letter code of the amino acid present in the mutation. If the second letter is 'X', any amino acid may replace the wildtype.
- Amino acid deletion - (e.g. ÄF508) The Greek symbol Ä or 'delta' indicates a deletion. The letter refers to the amino acid present in the wildtype and the number is the position from the N terminus of the amino acid were it to be present as in the wildtype.

Harmful Mutations

Changes in DNA caused by mutation can cause errors in protein sequence, creating partially or completely non-functional proteins. To function correctly, each cell depends on thousands of proteins to function in the right places at the right times. When a mutation alters a protein that plays a critical role in the body, a medical condition can result. A condition caused by mutations in one or more genes is called a genetic disorder. However, only a small percentage of mutations cause genetic disorders; most have no impact on health. For example, some mutations alter a gene's DNA base sequence but don't change the function of the protein made by the gene.

If a mutation is present in a germ cell, it can give rise to offspring that carries the mutation in all of its cells. This is the case in hereditary diseases. On the other hand, a mutation can occur in a somatic cell of an organism. Such mutations will be present in all descendants of this cell, and certain mutations can cause the cell to become malignant, and thus cause cancer.

Often, gene mutations that could cause a genetic disorder are repaired by the DNA repair system of the cell. Each cell has a number of pathways through which enzymes recognize and repair mistakes in DNA. Because DNA can be damaged or mutated in many ways, the process of DNA repair is an important way in which the body protects itself from disease.

Beneficial Mutations

A very small percentage of all mutations actually have a positive effect. These mutations lead to new versions of proteins that help an organism and its future generations better adapt to changes in their environment. For example, a specific 32 base pair deletion in human CCR5 (CCR5-32) confers HIV resistance to homozygotes and delays AIDS onset in heterozygotes. The CCR5 mutation is more common in those of European descent. One theory for the etiology of the relatively high frequency of CCR5-32 in the European population is that it conferred resistance to the bubonic plague in mid-14th century Europe.

Selection and Evolution

Mutations produce organisms with different genotypes, and those differences can result in different phenotypes. Many genetic mutations, called "neutral mutations", have a negligible effect on an organism's phenotype, health, and reproductive fitness. Mutations which do have an effect are often deleterious, but occassionally mutations arise which are beneficial in the current environmental context of the organism.

Population genetics research studies the distributions of these genetic differences within populations and how the distributions change over time. Changes in the frequency of an allele in a population can be influenced by natural selection, where a given allele's higher rate of survival and reproduction causes it to become more frequent in the population over time. Genetic drift can also occur, where chance events lead to random changes in allele frequency.

Over many generations, the genomes of organisms can change, resulting in the phenomenon of evolution. Mutations and the selection for beneficial mutations can cause a species to evolve into forms that better survive their environment, a process called adaptation. New species are formed through the process of speciation, a process often caused by geographical separations that allow different populations to genetically diverge.

As sequences diverge and change during the process of evolution, these differences between sequences can be used as a molecular clock to calculate the evolutionary distance between them. Genetic comparisons are generally considered the most accurate method of characterizing the relatedness between species, an improvement over the sometimes deceptive comparison of phenotypic characteristics. The evolutionary distances between species can be combined to form evolutionary trees. These trees are commonly considered the most accurate representation of relatedness, although the transfer of genetic material between unrelated species (known as "horizontal gene transfer" and most common in bacteria) cannot be represented by the tree.

CHROMOSOME

The total complement of genes in an organism or cell is known as its genome, which may be stored on one or more chromosomes; the region of the chromosome at which a particular gene is located is called its locus. A chromosome consists of a single, very long DNA helix on which thousands of genes are encoded. Prokaryotes - bacteria and archaea - typically store their genomes on a single large, circular chromosome, sometimes supplemented by additional small circles of DNA called plasmids, which usually encode only a few genes and are easily transferable between individuals. For example, the genes for antibiotic resistance are usually encoded on bacterial plasmids and can be passed between individual cells, even those of different species, via horizontal gene transfer. Although some simple eukaryotes also possess plasmids with small numbers of genes, the majority of eukaryotic genes are stored on multiple linear chromosomes, which are packed within the nucleus in complex with storage proteins called histones. The manner in which DNA is stored on the histone, as well as chemical modifications of the histone itself, are regulatory mechanisms governing whether a particular region of DNA is accessible for gene expression. The ends of eukaryotic chromosomes are capped by long stretches of repetitive sequences called telomeres, which do not code for any gene product but are present to prevent degradation of coding and regulatory regions during DNA replication. The length of the telomeres tends to decrease each time the genome is replicated in preparation for cell division; the loss of telomeres has been proposed as an explanation for cellular senescence, or the loss of the ability to divide, and by extension for the aging process in organisms.

While the chromosomes of prokaryotes are relatively gene-dense, those of eukaryotes often contain so-called "junk DNA", or regions of DNA that serve no obvious function. Simple single-celled eukaryotes have relatively small amounts of such DNA, while the genomes of complex multicellular organisms, including humans, contain an absolute majority of DNA without an identified function.

CHROMOSOMES AND GENDER

Chromosomes are long, stringy aggregates of genes that carry heredity information. They are composed of DNA and proteins and are located within the nucleus of our cells.

Chromosomes determine everything from hair colour and eye colour to gender. Whether you are a male or female depends on the presence or absence of certain chromosomes. Human cells contain 23 pairs of chromosomes for a total of 46. There are 22 pairs of autosomes and one pair of sex chromosomes. The sex chromosomes are the X chromosome and the Y chromosome. These chromosomes determine gender.

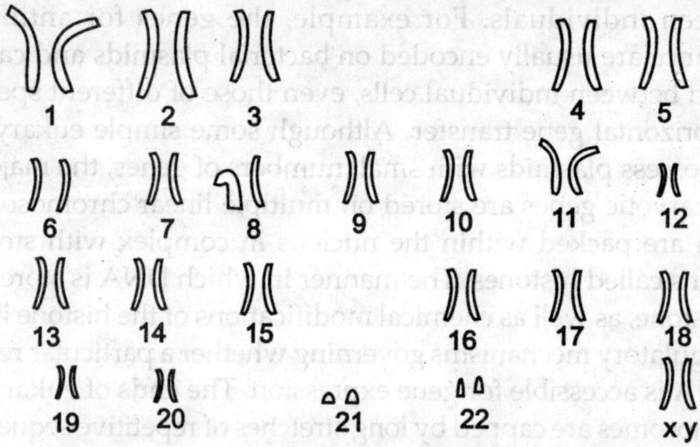

Fig. Karyotype of a normal male with 22 pairs of autosomes and one pair of sex chromosomes.

In human sexual reproduction, two distinct gametes fuse to form a zygote. Gametes are reproductive cells produced by a type of cell division called meiosis. They contain only one set of chromosomes and are said to be haploid (one set of 22 autosomes and one sex chromosome). The male gamete, called the spermatozoan, is relatively motile and usually has a flagellum. The female gamete, called the ovum, is nonmotile and relatively large in comparison to the male gamete. When the haploid male and female gametes unite in a process called fertilization, they form what is called a zygote. The zygote is diploid, meaning that it contains two sets of chromosomes (two sets of 22 autosomes and two sex chromosomes).

Sex Chromosomes X-Y

The male gametes or sperm cells in humans and other mammals are heterogametic and contain one of two types of sex chromosomes. They are either X or Y. The female gametes or eggs however, contain only the X sex chromosome and are homogametic. The sperm cell determines the sex of an individual in this case. If a sperm cell containing an X chromosome fertilizes an egg, the resulting zygote will be XX or female. If the sperm cell contains a Y chromosome, then the resulting zygote will be XY or male.

Sex Chromosomes X-O

Grasshoppers, roaches, and other insects have a similar system for determining the sex of an individual. Adult males lack a Y sex chromosome and have only an X chromosome. They produce sperm cells that contain either an X chromosome or no sex chromosome, which is designated as O. The females are XX and produce egg cells that contain an X chromosome. If an X sperm cell fertilizes an egg, the resulting zygote will be XX or female. If a sperm cell containing no sex chromosome fertilizes an egg, the resulting zygote will be XO or male.

Sex Chromosomes Z-W

Birds, insects like butterflies, and some species of fish have a different system for determining gender. In these animals it is the female gamete that determines the sex of an individual. Female gametes can either contain a Z chromosome or a W chromosome. Male gametes contain only the Z chromosome. Females of these species are ZW and males are ZZ.

Parthenogenesis

What about animals like most kinds of wasps, bees, and ants that have no sex chromosomes? How is gender determined? In these species, fertilization determines gender. If an egg becomes fertilized it will develop into a female. A non-fertilized egg may develop into a male. The female is diploid and contains two sets of chromosomes, while the male is haploid. This development of an unfertilized egg into an individual is called parthenogenesis.

Chapter 4

Working of Genes

People have known for many years that living things inherit traits from their parents. That common-sense observation led to agriculture, the purposeful breeding and cultivation of animals and plants for desirable characteristics. Firming up the details took quite some time, though. Researchers did not understand exactly how traits were passed to the next generation until the middle of the 20th century.

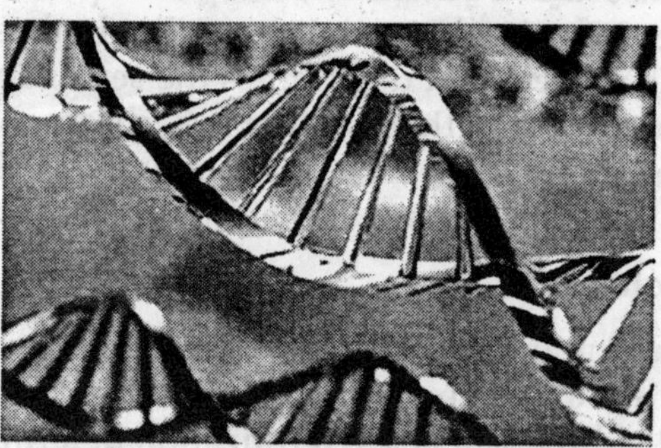

Fig. Structure of DNA.

Now it is clear that genes are what carry our traits through generations and that genes are made of deoxyribonucleic acid (DNA). But genes themselves don't do the actual work. Rather, they serve as instruction books for making functional molecules such as ribonucleic acid (RNA) and proteins, which perform the chemical reactions in our bodies.

Proteins do many other things, too. They provide the

body's main building materials, forming the cell's architecture and structural components. But one thing proteins can't do is make copies of themselves. When a cell needs more proteins, it uses the manufacturing instructions coded in DNA.

The DNA code of a gene—the sequence of its individual DNA building blocks, labeled A (adenine), T (thymine), C (cytosine), and G (guanine) and collectively called nucleotides—spells out the exact order of a protein's building blocks, amino acids.

Occasionally, there is a kind of typographical error in a gene's DNA sequence. This mistake— which can be a change, gap, or duplication—is called a mutation.

A mutation can cause a gene to encode a protein that works incorrectly or that doesn't work at all. Sometimes, the error means that no protein is made.

But not all DNA changes are harmful. Some mutations have no effect, and others produce new versions of proteins that may give a survival advantage to the organisms that have them. Over time, mutations supply the raw material from which new life forms evolve Beautiful DNA

Up until the 1950s, scientists knew a good deal about heredity, but they didn't have a clue what DNA looked like. In order to learn more about DNA and its structure, some scientists experimented with using X rays as a form of molecular photography.

In 1953, Watson and Crick created their historic model of the shape of DNA: the double helix.

Rosalind Franklin, a physical chemist working with Maurice Wilkins at King's College in London, was among the first to use this method to analyze genetic material. Her experiments produced what were referred to at the time as "the most beautiful X-ray photographs of any substance ever taken."

Other scientists, including zoologist James Watson and physicist Francis Crick, both working at Cambridge University in the United Kingdom, were trying to determine the shape of DNA too. Ultimately, this line of research revealed one of the most profound scientific discoveries of the 20th century: that DNA exists as a double helix.

The 1962 Nobel Prize in physiology or medicine was awarded to Watson, Crick, and Wilkins for this work. Although Franklin did not earn a share of the prize due to her untimely death at age 38, she is widely recognized as having played a significant role in the discovery.

The spiral staircase-shaped double helix has attained global status as the symbol for DNA. But what is so beautiful about the discovery of the twisting ladder structure isn't just its good looks. Rather, the structure of DNA taught researchers a fundamental lesson about genetics. It taught them that the two connected strands—winding together like parallel handrails—were complementary to each other, and this unlocked the secret of how genetic information is stored, transferred, and copied.

In genetics, complementary means that if you know the sequence of nucleotide building blocks on one strand, you know the sequence of nucleotide building blocks on the other strand: A always matches up with T and C always links to G.

Long strings of nucleotides form genes, and groups of genes are packaged tightly into structures called chromosomes. Every cell in your body except for eggs, sperm, and red blood cells contains a full set of chromosomes in its nucleus.

If the chromosomes in one of your cells were uncoiled and placed end to end, the DNA would be about 6 feet long. If all the DNA in your body were connected in this way, it would stretch approximately 67 billion miles! That's nearly 150,000 round trips to the Moon.

The long, stringy DNA that makes up genes is spooled within chromosomes inside the nucleus of a cell. (Note that a gene would actually be a much longer stretch of DNA than what is shown here.)

DNA consists of two long, twisted chains made up of nucleotides. Each nucleotide contains one base, one phosphate molecule, and the sugar molecule deoxyribose. The bases in DNA nucleotides are adenine, thymine, cytosine, and guanine.

Copycat

Fig. 23 pairs of chromosomes.

Male DNA (pictured here) contains an X and a Y chromosome, whereas female DNA contains two X chromosomes.

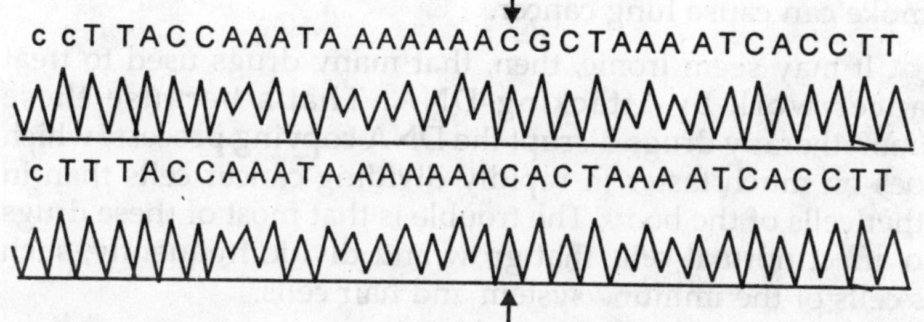

When DNA polymerase makes an error while copying a gene's DNA sequence, the mistake is called a mutation. In this example, the nucleotide G has been changed to an A.

It's astounding to think that your body consists of trillions of cells. But what's most amazing is that it all starts with one cell. How does this massive expansion take place?

As an embryo progresses through development, its cells must reproduce. But before a cell divides into two new, nearly identical cells, it must copy its DNA so there will be a complete set of genes to pass on to each of the new cells.

To make a copy of itself, the twisted, compacted double helix of DNA has to unwind and separate its two strands. Each strand becomes a pattern, or template, for making a new strand, so the two new DNA molecules have one new strand and one old strand.

The copy is courtesy of a cellular protein machine called DNA polymerase, which reads the template DNA strand and stitches together the complementary new strand. The process, called replication, is astonishingly fast and accurate, although occasional mistakes, such as deletions or duplications, occur. Fortunately, a cellular spell-checker catches and corrects nearly all of these errors.

Mistakes that are not corrected can lead to diseases such as cancer and certain genetic disorders. Some of these include Fanconi anemia, early aging diseases, and other conditions in which people are extremely sensitive to sunlight and some chemicals.

DNA copying is not the only time when DNA damage can happen. Prolonged, unprotected sun exposure can cause DNA changes that lead to skin cancer, and toxins in cigarette smoke can cause lung cancer.

It may seem ironic, then, that many drugs used to treat cancer work by attacking DNA. That's because these chemotherapy drugs disrupt the DNA copying process, which goes on much faster in rapidly dividing cancer cells than in other cells of the body. The trouble is that most of these drugs do affect normal cells that grow and divide frequently, such as cells of the immune system and hair cells.

Understanding DNA replication better could be a key to limiting a drug's action to cancer cells only.

LET'S CALL IT EVEN

After copying its DNA, a cell's next challenge is getting just the right amount of genetic material into each of its two offspring.

Most of your cells are called diploid ("di" means two, and "ploid" refers to sets of chromosomes) because they have two sets of chromosomes (23 pairs). Eggs and sperm are different; these are known as haploid cells. Each haploid cell has only one set of 23 chromosomes so that at fertilization the math will work out: A haploid egg cell will combine with a haploid sperm cell to form a diploid cell with the right number of chromosomes: 46.

Chromosomes are numbered 1 to 22, according to size, with 1 being the largest chromosome. The 23rd pair, known as the sex chromosomes, are called X and Y. In humans, abnormalities of chromosome number usually occur during meiosis, the time when a cell reduces its chromosomes from diploid to haploid in creating eggs or sperm.

What happens if an egg or a sperm cell gets the wrong number of chromosomes, and how often does this happen?

Molecular biologist Angelika Amon of the Massachusetts Institute of Technology in Cambridge says that mistakes in dividing DNA between daughter cells during meiosis are the leading cause of human birth defects and miscarriages. Current estimates are that 10 percent of all embryos have an incorrect chromosome number. Most of these don't go to full term and are miscarried.

In women, the likelihood that chromosomes won't be apportioned properly increases with age. One of every 18 babies born to women over 45 has three copies of chromosome 13, 18, or 21 instead of the normal two, and this improper balancing can cause trouble. For example, three copies of chromosome 21 lead to Down syndrome.

To make her work easier, Amon—like many other basic scientists—studies yeast cells, which separate their chromosomes almost exactly the same way human cells do, except that yeast do it much faster. A yeast cell copies its DNA and produces daughter cells in about 11/2 hours, compared to a whole day for human cells.

The yeast cells she uses are the same kind bakeries use to make bread and breweries use to make beer!

Amon has made major progress in understanding the details of meiosis. Her research shows how, in healthy cells, gluelike protein complexes called cohesins release pairs of chromosomes at exactly the right time. This allows the chromosomes to separate properly.

These findings have important implications for understanding and treating infertility, birth defects, and cancer.

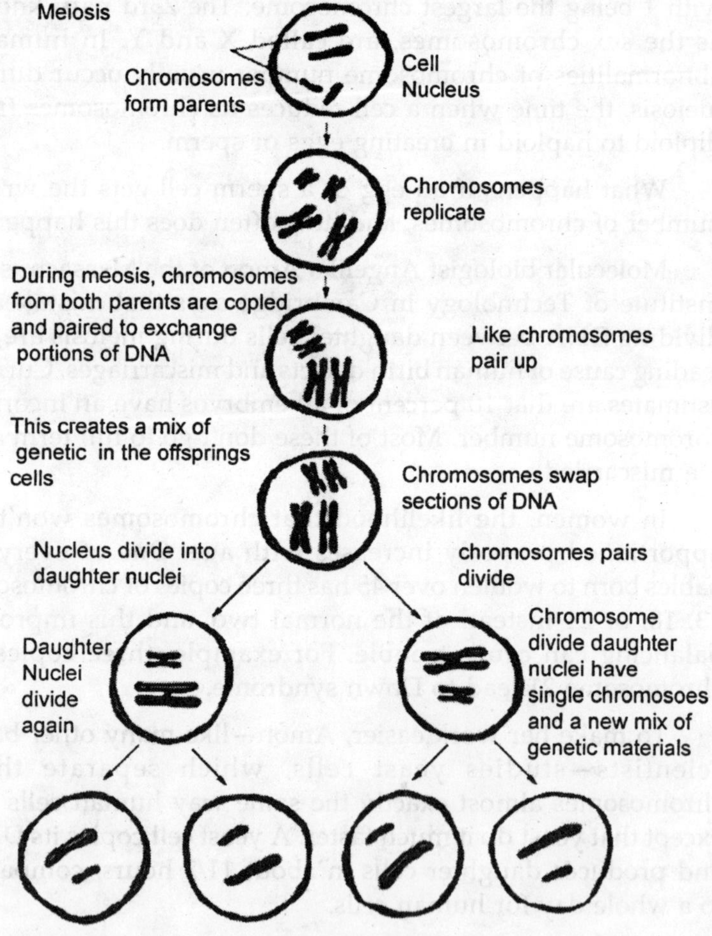

Fig .meiosis

GETTING THE MESSAGE

So, we've described DNA—its basic properties and how our bodies make more of it. But how does DNA serve as the language of life? How do you get a protein from a gene?

There are two major steps in making a protein. The first is transcription, where the information coded in DNA is copied into RNA. The RNA nucleotides are complementary to those on the DNA: a C on the RNA strand matches a G on the DNA strand.

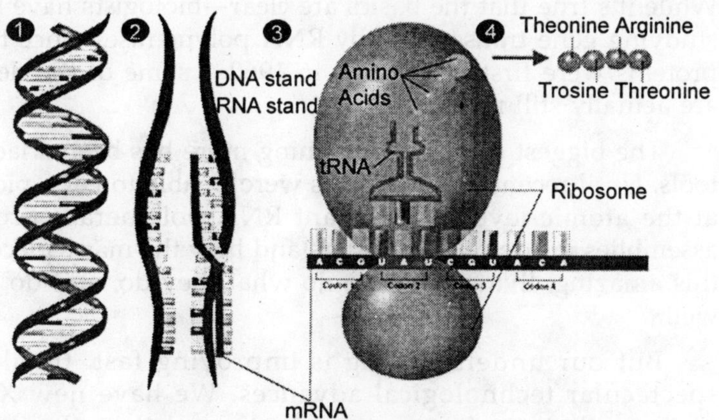

Fig. RNA polymerase transcribes DNA

RNA polymerase transcribes DNA to make messenger RNA (mRNA). The mRNA sequence (dark red strand) is complementary to the DNA sequence (blue strand). On ribosomes, transfer RNA (tRNA) helps convert mRNA into protein. Amino acids link up to make a protein.

The only difference is that RNA pairs a nucleotide called uracil (U), instead of a T, with an A on the DNA.

A protein machine called RNA polymerase reads the DNA and makes the RNA copy. This copy is called messenger RNA, or mRNA, because it delivers the gene's message to the protein-producing machinery.

At this point you may be wondering why all of the cells in the human body aren't exactly alike, since they all contain the same DNA.What makes a liver cell different from a brain cell? How do the cells in the heart make the organ contract, but those in skin allow us to sweat?

That's because RNA polymerase does not work alone, but rather functions with the aid of many helper proteins. While the core part of RNA polymerase is the same in all cells, the helpers vary in different cell types throughout the body.

You'd think that for a process so essential to life, researchers would know a lot about how transcription works. While it's true that the basics are clear—biologists have been studying gene transcribing by RNA polymerases since these proteins were first discovered in 1960— some of the details are actually still murky.

The biggest obstacle to learning more has been a lack of tools. Until recently, researchers were unable to get a picture at the atomic level of the giant RNA polymerase protein assemblies inside cells to understand how the many pieces of this amazing, living machine do what they do, and do it so well.

But our understanding is improving fast, thanks to spectacular technological advances. We have new X-ray pictures that are far more sophisticated than those that revealed the structure of DNA. Roger Kornberg of Stanford University in California used such methods to determine the structure of RNA polymerase. This work earned him the 2006 Nobel Prize in chemistry. In addition, very powerful microscopes and other tools that allow us to watch one molecule at a time provide a new look at RNA polymerase while it's at work reading DNA and producing RNA.

For example, Steven Block, also of Stanford, has used a physics technique called optical trapping to track RNA polymerase as it inches along DNA. Block and his team performed this work by designing a specialized microscope sensitive enough to watch the real-time motion of a single polymerase traveling down a gene on one chromosome.

The researchers discovered that molecules of RNA polymerase behave like battery-powered spiders as they crawl along the DNA ladder, adding nucleotides one at a time to the growing RNA strand. The enzyme works much like a motor, Block believes, powered by energy released during the chemical synthesis of DNA.

NATURE'S CUT-AND-PASTE JOB

Genes are often interrupted by stretches of DNA (introns, blue) that do not contain instructions for making a protein. The DNA segments that do contain protein-making instructions are known as exons (green).

Several types of RNA play key roles in making a protein. The gene transcript (the mRNA) transfers information from DNA in the nucleus to the ribosomes that make protein. Ribosomal RNA forms about 60 percent of the ribosomes. Lastly, transfer RNA carries amino acids to the ribosomes. As you can see, all three types of cellular RNAs come together to produce new proteins.

But the journey from gene to protein isn't quite as simple as we've just made it out to be. After transcription, several things need to happen to mRNA before a protein can be made. For example, the genetic material of humans and other eukaryotes (organisms that have a nucleus) includes a lot of DNA that doesn't encode proteins. Some of this DNA is stuck right in the middle of genes.

To distinguish the two types of DNA, scientists call the coding sequences of genes exons and the pieces in between introns (for intervening sequences).

If RNA polymerase were to transcribe DNA from the start of an intron-containing gene to the end, the RNA would be complementary to the introns as well as the exons.

To get an mRNA molecule that yields a working protein, the cell needs to trim out the intron sections and then stitch only the exon pieces together. This process is called RNA splicing.

Arranging exons in different patterns, called alternative splicing, enables cells to make different proteins from a single gene.

Splicing has to be extremely accurate. An error in the splicing process, even one that results in the deletion of just one nucleotide in an exon or the addition of just one nucleotide in an intron, will throw the whole sequence out of alignment. The result is usually an abnormal protein—or no protein at all. One form of Alzheimer's disease, for example, is caused by this kind of splicing error.

Molecular biologist Christine Guthrie of the University of California, San Francisco, wants to understand more fully the mechanism for removing intron RNA and find out how it stays so accurate.

She uses yeast cells for these experiments. Just like human DNA, yeast DNA has introns, but they are fewer and simpler in structure and are therefore easier to study. Guthrie can identify which genes are required for splicing by finding abnormal yeast cells that mangle splicing.

So why do introns exist, if they're just going to be chopped out? Without introns, cells wouldn't need to go through the splicing process and keep monitoring it to be sure it's working right.

As it turns out, splicing also makes it possible for cells to create more proteins.

Think about all the exons in a gene. If a cell stitches together exons 1, 2, and 4, leaving out exon 3, the mRNA will specify the production of a particular protein. But instead, if the cell stitches together exons 1, 2, and 3, this time leaving out exon 4, then the mRNA will be translated into a different protein.

By cutting and pasting the exons in different patterns, which scientists call alternative splicing, a cell can create different proteins from a single gene. Alternative splicing is one of the reasons why human cells, which have about 25,000 genes, can make hundreds of thousands of different proteins.

ALL TOGETHER NOW

Until recently, researchers looked at genes, and the proteins they encode, one at a time. Now, they can look at how large numbers of genes and proteins act, as well as how they interact. This gives them a much better picture of what goes on in a living organism.

Already, scientists can identify all of the genes that are transcribed in a cell—or in an organ, like the heart. And although researchers can't tell you, right now, what's going on in every cell of your body while you read a book or walk down the street, they can do this sort of "whole-body" scan for simpler, single-celled organisms like yeast.

Using a new technique called genome-wide location analysis, Richard Young of the Massachusetts Institute of Technology unraveled a "regulatory code" of living yeast cells, which have more than 6,000 genes in their genome. Young's technique enabled him to determine the exact places where RNA polymerase's helper proteins sit on DNA and tell RNA polymerase to begin transcribing a gene.

Since he did the experiment with the yeast exposed to a variety of different conditions, Young was able to Fig out how transcription patterns differ when the yeast cell is under stress (say, in a dry environment) or thriving in a sugary-rich nutrient solution. Done one gene at a time, using methods considered state-of-the-art just a few years ago, this kind of analysis would have taken hundreds of years.

After demonstrating that his technique worked in yeast, Young then took his research a step forward. He used a variation of the yeast method to scan the entire human genome in small samples of cells taken from the pancreases and livers of people with type 2 diabetes. He used the results to identify genes that aren't transcribed correctly in people with the disease.

This information provides researchers with an important tool for understanding how diabetes and other diseases are influenced by defective genes. By building models to predict how genes respond in diverse situations, researchers may be able to learn how to stop or jump-start genes on demand, change the course of a disease, or prevent it from ever

NURSERY GENETICS

While most genetic research uses lab organisms, test tubes, and petri dishes, the results have real consequences for people. Your first encounter with genetic analysis probably happened shortly after you were born, when a doctor or nurse took a drop of blood from the heel of your tiny foot.

Lab tests performed with that single drop of blood can diagnose certain rare genetic disorders as well as metabolic problems like phenylketonuria (PKU).

Screening newborns in this way began in the 1960s in Massachusetts with testing for PKU, a disease affecting 1 in 14,000 people. PKU is caused by an enzyme that doesn't work properly due to a genetic mutation. Those born with this disorder cannot metabolize the amino acid phenylalanine, which is present in many foods. Left untreated, PKU can lead to mental retardation and neurological damage, but a special diet can prevent these outcomes. Testing for this condition has made a huge difference in many lives.

Newborn screening is governed by individual states. That means that the state in which a baby is born determines the genetic conditions for which he or she will be screened. Currently, some states test for fewer than 10 conditions, whereas others test for more than 30. All states test for PKU.

Although expanded screening for genetic diseases in newborns is advocated by some, others question the value of screening for conditions that are currently untreatable. Another issue is that some children with mild versions of certain genetic diseases may be treated needlessly.

Based on recommendations released in 2005 by the American College of Medical Genetics, government policymakers are now considering a proposal that would establish a standard, national set of newborn tests for 29 conditions, ranging from relatively common hearing problems to very rare metabolic diseases.

TRANSLATION

After a gene has been read by RNA polymerase and the RNA is spliced, what happens next in the journey from gene to protein? The next step is reading the RNA information and fitting the building blocks of a protein together. This is called translation, and its principal actors are the ribosome and amino acids.

Ribosomes are among the biggest and most intricate structures in the cell. The ribosomes of bacteria contain not only huge amounts of RNA, but also more than 50 different proteins. Human ribosomes have even more RNA and between 70 and 80 different proteins!

Harry Noller of the University of California, Santa Cruz, has found that a ribosome performs several key jobs when it translates the genetic code of mRNA. As the messenger RNA threads through the ribosome protein machine, the ribosome reads the mRNA sequence and helps recognize and recruit the correct amino acid-carrying transfer RNA to match the mRNA code. The ribosome also links each additional amino acid into a growing protein chain.

For many years, researchers believed that even though RNAs formed a part of the ribosome, the protein portion of the ribosome did all of the work. Noller thought, instead, that maybe RNA, not proteins, performed the ribosome's job. His idea was not popular at first, because at that time it was thought that RNA could not perform such complex functions.

Some time later, however, the consensus changed. Sidney Altman of Yale University in New Haven, Connecticut, and Thomas Cech, who was then at the University of Colorado in Boulder, each discovered that RNA can perform work as complex as that done by protein enzymes. Their "RNA-as-an-enzyme" discovery turned the research world on its head and earned Cech and Altman the 1989 Nobel Prize in chemistry.

Noller and other researchers have continued the painstaking work of understanding ribosomes. In 1999, he showed how different parts of a bacterial ribosome interact with one another and how the ribosome interacts with molecules involved in protein synthesis. These studies provided near proof that the fundamental mechanism of translation is performed by RNA, not by the proteins of the ribosome.

RNA SURPRISES

But which ribosomal RNAs are doing the work? Most scientists assumed that RNA nucleotides buried deep within the ribosome complex—the ones that have the same sequence in every species from bacteria to people—were the important ones for piecing the growing protein together.

However, recent research by Rachel Green, who worked with Noller before moving to Johns Hopkins University in Baltimore, Maryland, showed that this is not the case. Green discovered that those RNA nucleotides are not needed for assembling a protein. Instead, she found, the nucleotides do something else entirely: They help the growing protein slip off the ribosome once it's finished.

Noller, Green, and hundreds of other scientists work with the ribosomes of bacteria. Why should you care about how bacteria create proteins from their genes?

One reason is that this knowledge is important for learning how to disrupt the actions of disease-causing microorganisms. For example, antibiotics like erythromycin and neomycin work by attacking the ribosomes of bacteria, which are different enough from human ribosomes that our cells are not affected by these drugs.

As researchers gain new information about bacterial translation, the knowledge may lead to more antibiotics for people.

New antibiotics are urgently needed because many bacteria have developed resistance to the current arsenal. This resistance is sometimes the result of changes in the bacteria's ribosomal RNA. It can be difficult to find those small, but critical, changes that may lead to resistance, so it is important to find completely new ways to block bacterial translation.

Green is working on that problem too. Her strategy is to make random mutations to the genes in a bacterium that affect its ribosomes. But what if the mutation disables the ribosome so much that it can't make proteins? Then the bacterium won't grow, and Green wouldn't find it.

Using clever molecular tricks, Green Figd out a way to rescue some of the bacteria with defective ribosomes so they could grow. While some of the rescued bacteria have changes in their ribosomal RNA that make them resistant to certain antibiotics (and thus would not make good antibiotic targets) other RNA changes that don't affect resistance may point to promising ideas for new antibiotics.

AN INTERESTING DEVELOPMENT

In the human body, one of the most important jobs for proteins is to control how embryos develop. Scientists discovered a hugely important set of proteins involved in development by studying mutations that cause bizarre malformations in fruit flies.

The most famous such abnormality is a fruit fly with a leg, rather than the usual antenna, growing out of its head. According to Thomas C. Kaufman of Indiana University in Bloomington, the leg is perfectly normal—it's just growing in the wrong place.

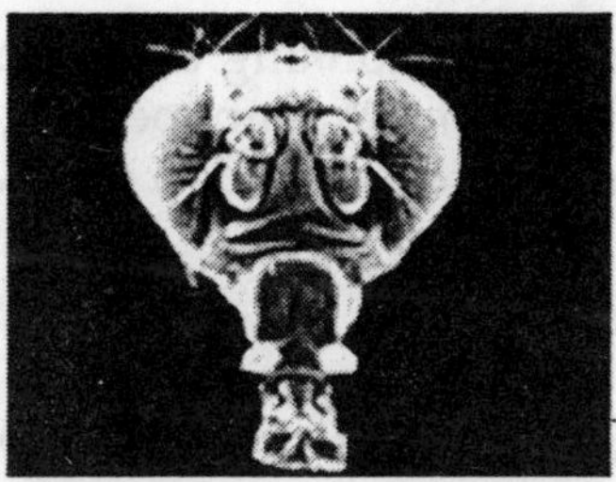

Fig. Normal fruit fly head.

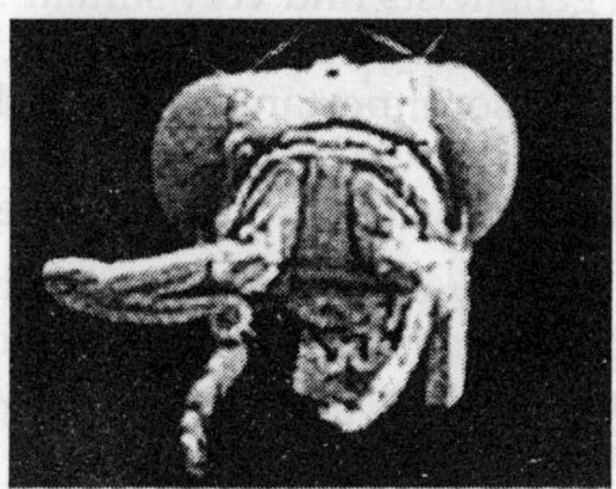

Fig. Fruit fly head showing the effects of the Antennapedia gene. This fly has legs where its antennae should be.

In this type of mutation and many others, something goes wrong with the genetic programme that directs some of the cells in an embryo to follow developmental pathways, which are a series of chemical reactions that occur in a specific order. In the antenna-into-leg problem, it is as if the cells growing from the fly's head, which normally would become an antenna, mistakenly believe that they are in the fly's thorax, and therefore ought to grow into a leg. And so they do.

Thinking about this odd situation taught scientists an important lesson—that the proteins made by some genes can act as switches. Switch genes are master controllers that provide each body part with a kind of identification card. If a protein that normally instructs cells to become an antenna is disrupted, cells can receive new instructions to become a leg instead.

Scientists determined that several different genes, each with a common sequence, provide these anatomical identification card instructions. Kaufman isolated and described one of these genes, which became known as Antennapedia, a word that means "antenna feet."

Kaufman then began looking a lot more closely at the molecular structure of the Antennapedia gene. In the early 1980s, he and other researchers made a discovery that has been fundamental to understanding evolution as well as developmental biology.

The scientists found a short sequence of DNA, now called the homeobox, that is present not only in Antennapedia but in the several genes next to it and in genes in many other organisms.When geneticists find very similar DNA sequences in the genes of different organisms, it's a good clue that these genes do something so important and useful that evolution uses the same sequence over and over and permits very few changes in its structure as new species evolve.

Researchers quickly discovered nearly identical versions of homeobox DNA in almost every non-bacterial cell they examined—from yeast to plants, frogs, worms, beetles, chickens, mice, and people.

Hundreds of homeobox-containing genes have been identified, and the proteins they make turn out to be involved in the early stages of development of many species. For example, researchers have found that abnormalities in the homeobox genes can lead to extra fingers or toes in humans.

RNA AND DNA REVEALED: NEW ROLES, NEW RULES

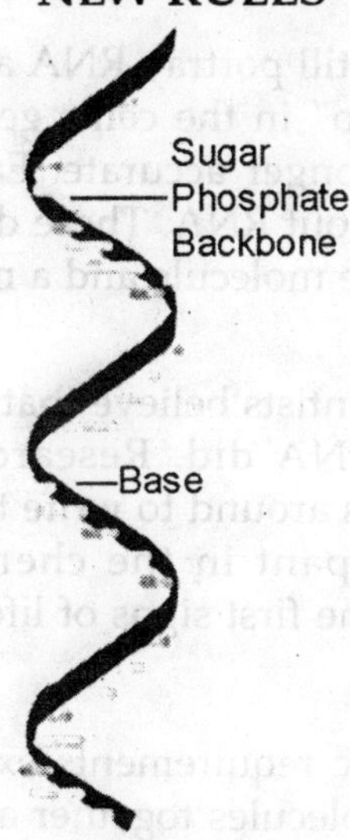

Fig. Ribonucleic acid (RNA) has the bases adenine (A), cytosine (C), guanine (G), and uracil (U).

For many years, when scientists thought about heredity, DNA was the first thing to come to mind. It's true that DNA is the basic ingredient of our genes and, as such, it often steals the limelight from RNA, the other form of genetic material inside our cells.

But, while they are both types of genetic material, RNA and DNA are rather different. The chemical units of RNA are like those of DNA, except that RNA has the nucleotide uracil (U) instead of thymine (T). Unlike double-stranded DNA, RNA usually comes as only a single strand. And the nucleotides in RNA contain ribose sugar molecules in place of deoxyribose.

RNA is quite flexible—unlike DNA, which is a rigid, spiral-staircase molecule that is very stable. RNA can twist itself into a variety of complicated, three-dimensional shapes. RNA is also unstable in that cells constantly break it down and must continually make it fresh, while DNA is not broken down often. RNA's instability lets cells change their patterns of protein synthesis very quickly in response to what's going on around them.

Many textbooks still portray RNA as a passive molecule, simply a "middle step" in the cell's gene-reading activities. But that view is no longer accurate. Each year, researchers unlock new secrets about RNA. These discoveries reveal that it is truly a remarkable molecule and a multi-talented actor in heredity.

Today, many scientists believe that RNA evolved on the Earth long before DNA did. Researchers hypothesize—obviously, no one was around to write this down—that RNA was a major participant in the chemical reactions that ultimately spawned the first signs of life on the planet.

RNA WORLD

At least two basic requirements exist for making a cell: the ability to hook molecules together and break them apart, and the ability to replicate, or copy itself, from existing information.

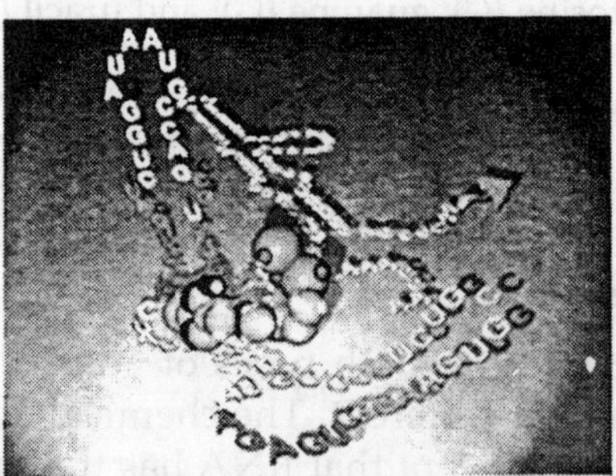

Fig.Riboswitches

Riboswitches are RNA sequences that control gene activity. The riboswitch shown here bends into a special shape when it grips tightly onto a molecule called a metabolite (colored balls) that bacteria need to survive.

RNA probably helped to form the first cell. The first organic molecules, meaning molecules containing carbon, most likely arose out of random collisions of gases in the Earth's primitive atmosphere, energy from the Sun, and heat from naturally occurring radioactivity. Some scientists think that in this primitive world, RNA was a critical molecule because of its ability to lead a double life: to store information and to conduct chemical reactions. In other words, in this world, RNA served the functions of both DNA and proteins.

What does any of this have to do with human health? Plenty, it turns out.

Today's researchers are harnessing some of RNA's flexibility and power. For example, through a strategy he calls directed evolution, molecular engineer Ronald R. Breaker of Yale University is developing ways to create entirely new forms of RNA and DNA that both work as enzymes.

Recently, Breaker and others have also uncovered a hidden world of RNAs that play a major role in controlling gene activity, a job once thought to be performed exclusively by proteins. These RNAs, which the scientists named riboswitches, are found in a wide variety of bacteria and other organisms.

This discovery has led Breaker to speculate that new kinds of antibiotic medicines could be developed to target bacterial riboswitches.

Molecular Editor

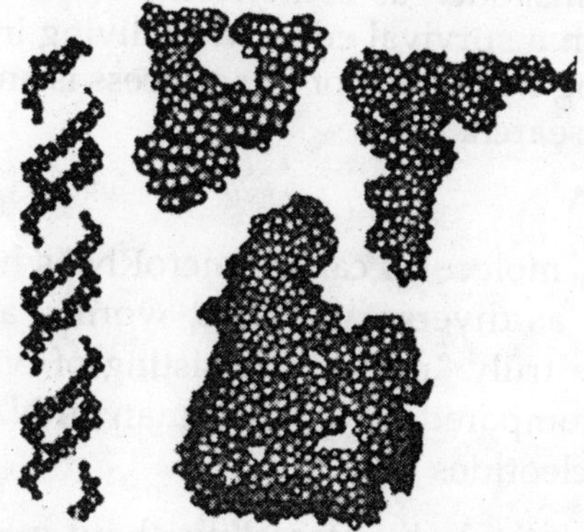

Fig.RNA comes in a variety of different shapes. Double-stranded DNA is a staircase-like molecule.

Scientists are learning of another way to customise proteins: by RNA editing. Although DNA sequences spell out instructions for producing RNA and proteins, these instructions aren't always followed precisely. Editing a gene's mRNA, even by a single chemical letter, can radically change the resulting protein's function. Nature likely evolved the RNA editing function as a way to get more proteins out of the same number of genes. For example, researchers have found that the mRNAs for certain proteins important for the proper functioning of the nervous system are particularly prone to editing. It may be that RNA editing gives certain brain cells the capacity to react quickly to a changing environment.

Which molecules serve as the editor and how does this happen? Brenda Bass of the University of Utah School of Medicine in Salt Lake City studies one particular class of editors called adenosine deaminases. These enzymes "retype" RNA letters at various places within an mRNA transcript.

They do their job by searching for characteristic RNA shapes. Telltale twists and bends in folded RNA molecules signal these enzymes to change the RNA sequence, which in turn changes the protein that gets made.

Bass' experiments show that RNA editing occurs in a variety of organisms, including people. Another interesting aspect of editing is that certain disease-causing microorganisms, such as some forms of parasites, use RNA editing to gain a survival edge when living in a human host. Understanding the details of this process is an important area of medical research.

MICRO RNA

Recently, molecules called microRNAs have been found in organisms as diverse as plants, worms, and people. The molecules are truly "micro," consisting of only a few dozen nucleotides, compared to typical human mRNAs that are a few thousand nucleotides long.

What's particularly interesting about microRNAs is that many of them arise from DNA that used to be considered merely filler material.

How do these small but important RNA molecules do their work? They start out much bigger but get trimmed by cellular enzymes, including one aptly named Dicer. Like tiny pieces of Velcro®, microRNAs stick to certain mRNA molecules and stop them from passing on their protein-making instructions.

First discovered in a roundworm model system some microRNAs help determine the organism's body plan. In their absence, very bad things can happen. For example, worms engineered to lack a microRNA called let-7 develop so abnormally that they often rupture and practically break in half as the worm grows.

Perhaps it is not surprising that since microRNAs help specify the timing of an organism's developmental plan, the appearance of the microRNAs themselves is carefully timed inside a developing organism. Biologists, including Amy Pasquinelli of the University of California, San Diego, are currently figuring out how microRNAs are made and cut to size, as well as how they are produced at the proper time during development.

MicroRNA molecules also have been linked to cancer. For example, Gregory Hannon of the Cold Spring Harbor Laboratory on Long Island, New York, found that certain microRNAs are associated with the severity of the blood cancer B-cell lymphoma in mice.

Since the discovery of microRNAs in the first years of the 21st century, scientists have identified hundreds of them that likely exist as part of a large family with similar nucleotide sequences. New roles for these molecules are still being found.

Healthy Interference

RNA controls genes in a way that was only discovered recently: a process called RNA interference, or RNAi. Although scientists identified RNAi less than 10 years ago, they now know that organisms have been using this trick for millions of years.

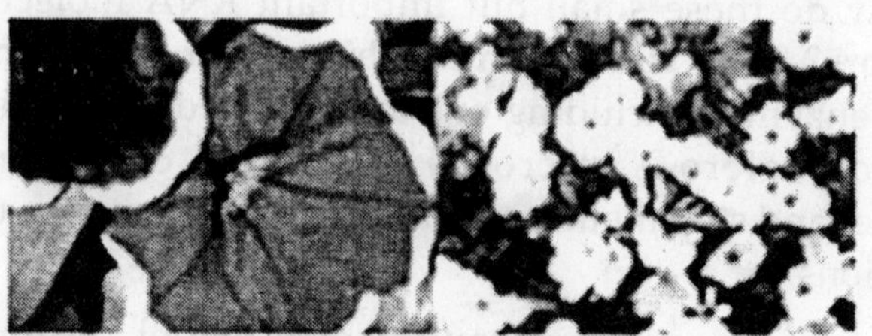

Fig. Gene Encoded

Researchers believe that RNAi arose as a way to reduce the production of a gene's encoded protein for purposes of fine-tuning growth or self-defence. When viruses infect cells, for example, they command their host to produce specialized RNAs that allow the virus to survive and make copies of itself. Researchers believe that RNAi eliminates unwanted viral RNA, and some speculate that it may even play a role in human immunity.

Oddly enough, scientists discovered RNAi from a failed experiment! Researchers investigating genes involved in plant growth noticed something strange: When they tried to turn petunia flowers purple by adding an extra "purple" gene, the flowers bloomed white instead.

This result fascinated researchers, who could not understand how adding genetic material could somehow get rid of an inherited trait. The mystery remained unsolved until, a few years later, two geneticists studying development saw a similar thing happening in lab animals.

The researchers, Andrew Z. Fire, then of the Carnegie Institution of Washington in Baltimore and now at Stanford University, and Craig Mello of the University of Massachusetts Medical School in Worcester, were trying to block the expression of genes that affect cell growth and tissue formation in roundworms, using a molecular tool called antisense RNA.

To their surprise, Mello and Fire found that their antisense RNA tool wasn't doing much at all. Rather, they determined, a doublestranded contaminant produced during the synthesis of the single-stranded antisense RNA interfered with gene expression. Mello and Fire named the process RNAi, and in 2006 were awarded the Nobel Prize in physiology or medicine for their discovery.

Further experiments revealed that the doublestranded RNA gets chopped up inside the cell into much smaller pieces that stick to mRNA and block its action, much like the microRNA pieces of Velcro discussed above.

Today, scientists are taking a cue from nature and using RNAi to explore biology. They have learned, for example, that the process is not limited to worms and plants, but operates in humans too.

Medical researchers are currently testing new types of RNAi-based drugs for treating conditions such as macular degeneration, the leading cause of blindness, and various infections, including those caused by HIV and herpes virus.

Dynamic DNA

A good part of who we are is "written in our genes," inherited from Mom and Dad. Many traits, like red or brown hair, body shape, and even some personality quirks, are passed on from parent to offspring.

But genes are not the whole story. Where we live, how much we exercise, what we eat: These and many other environmental factors can all affect how our genes get expressed.

You know that changes in DNA and RNA can produce changes in proteins. But additional control happens at the level of DNA, even though these changes do not alter DNA directly. Inherited factors that do not change the DNA sequence of nucleotides are called epigenetic changes, and they too help make each of us unique.

Epigenetic means, literally, "upon" or "over" genetics. It describes a type of chemical reaction that can alter the physical properties of DNA without changing its sequence. These changes make genes either more or less likely to be expressed.

Currently, scientists are following an intriguing course of discovery to identify epigenetic factors that, along with diet and other environmental influences, affect who we are and what type of illnesses we might get.

Secret Code

DNA is spooled up compactly inside cells in an arrangement called chromatin. This packaging is critical for DNA to do its work. Chromatin consists of long strings of DNA spooled around a compact assembly of proteins called histones.

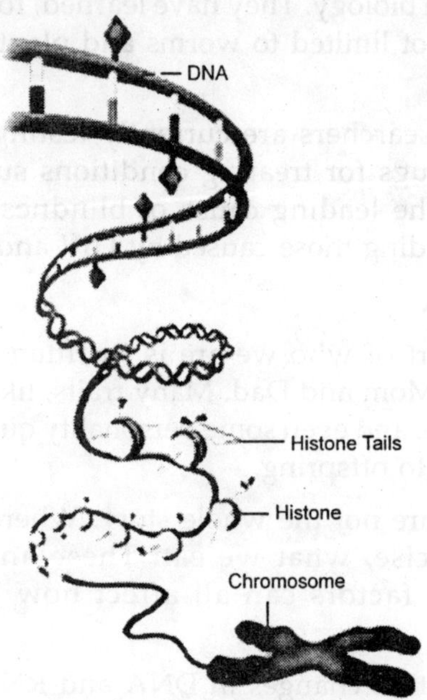

Fig. The epigenetic code

The "epigenetic code" controls gene activity with chemical tags that mark DNA (purple diamonds) and the "tails" of histone proteins (purple triangles). These markings help determine whether genes will be transcribed by RNA polymerase. Genes hidden from access to RNA polymerase are not expressed.

One of the key functions of chromatin is to control access to genes, since not all genes are turned on at the same time. Improper expression of growth-promoting genes, for example, can lead to cancer, birth defects, or other health concerns.

Many years after the structure of DNA was determined, researchers used a powerful device known as an electron microscope to take pictures of chromatin fibers. Upon viewing chromatin up close, the researchers described it as "beads on a string," an image still used today. The beads were the histone balls, and the string was DNA wrapped around the histones and connecting one bead to the next.

Decades of study eventually revealed that histones have special chemical tags that act like switches to control access to the DNA. Flipping these switches, called epigenetic markings, unwinds the spooled DNA so the genes can be transcribed.

The observation that a cell's gene-reading machinery tracks epigenetic markings led C. David Allis, who was then at the University of Virginia Health Sciences Centre in Charlottesville and now works at the Rockefeller University in New York City, to coin a new phrase, the "histone code." He and others believe that the histone code plays a major role in determining which proteins get made in a cell.

Flaws in the histone code have been associated with several types of cancer, and researchers are actively pursuing the development of medicines to correct such errors.

BATTLE OF THE SEXES

A process called imprinting, which occurs naturally in our cells, provides another example of how epigenetics affects gene activity.

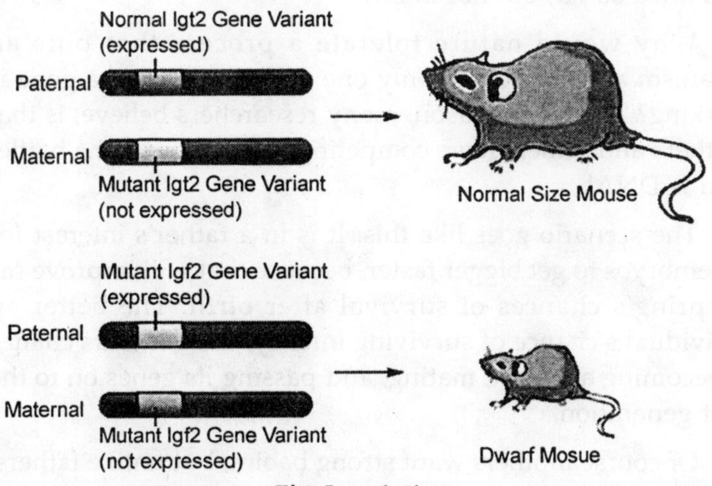

Fig. Imprinting

Igf2 is an imprinted gene. A single copy of the abnormal, or mutant, form of the Igf2 gene (red) causes growth defects, but only if the abnormal gene variant is inherited from the father.

With most genes, the two copies work exactly the same way. For some mammalian genes, however, only the mother's or the father's copy is switched on regardless of the child's gender. This is because the genes are chemically marked, or imprinted, during the process that generates eggs and sperm.

As a result, the embryo that emerges from the joining of egg and sperm can tell whether a gene copy came from Mom or Dad, so it knows which copy of the gene to shut off.

One example of an imprinted gene is insulin-like growth factor 2 (Igf2), a gene that helps a mammalian fetus grow. In this case, only the father's copy of Igf2 is expressed, and the mother's copy remains silent (is not expressed) throughout the life of the offspring.

Scientists have discovered that this selective silencing of Igf2 and many other imprinted genes occurs in all placental mammals (all except the platypus, echidna, and marsupials) examined so far, but not in birds.

Why would nature tolerate a process that puts an organism at risk because only one of two copies of a gene is working? The likely reason, many researchers believe, is that mothers and fathers have competing interests, and the battlefield is DNA!

The scenario goes like this: It is in a father's interest for his embryos to get bigger faster, because that will improve his offspring's chances of survival after birth. The better an individual's chance of surviving infancy, the better its chance of becoming an adult, mating, and passing its genes on to the next generation.

Of course mothers want strong babies, but unlike fathers, mothers provide physical resources to embryos during pregnancy. Over her lifetime, a female is likely to be pregnant several times, so she needs to divide her resources among a number of embryos in different pregnancies.

Researchers have discovered over 200 imprinted genes in mammals since the first one was identified in 1991. We now know that imprinting controls some of the genes that have an important role in regulating embryonic and fetal growth and allocating maternal resources. Not surprisingly, mutations in these genes cause serious growth disorders.

Marisa Bartolomei of the University of Pennsylvania School of Medicine in Philadelphia is trying to Fig out how Igf2 and other genes become imprinted and stay silent throughout the life of an individual. She has already identified sequences within genes that are essential for imprinting. Bartolomei and other researchers have shown that these sequences, called insulators, serve as "landing sites" for a protein that keeps the imprinted gene from being transcribed.

Starting at the End

When we think of DNA, we think of genes. However, some DNA sequences are different: They don't encode RNAs or proteins. Introns, described earlier are in this category.

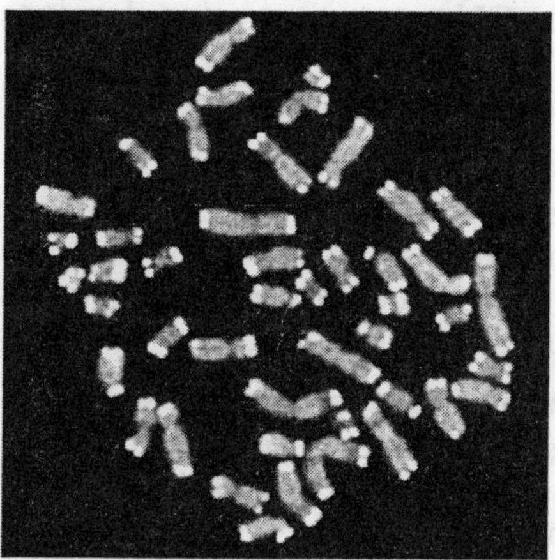

Fig. Telomeres, repeated nucleotide sequences at the tips of chromosomes, appear white in this figure.

Another example is telomeres—the ends of chromosomes. There are no genes in telomeres, but they serve an essential function. Like shoelaces without their tips, chromosomes without telomeres unravel and fray. And without telomeres, chromosomes stick to each other and cause cells to undergo harmful changes like dividing abnormally.

Researchers know a good deal about telomeres, dating back to experiments performed in the 1970s by Elizabeth Blackburn, a basic researcher who was curious about some of the fundamental events that take place within cells.

At the time, Blackburn, now at the University of California, San Francisco, was working with Joseph Gall at Yale University. For her experimental system, she chose a single-celled, pond-dwelling organism named Tetrahymena. These tiny, pear-shaped creatures are covered with hairlike cilia that they use to propel themselves through the water as they devour bacteria and fungi.

Tetrahymena was a good organism for Blackburn's experiments because it has a large number of chromosomes—which means it has a lot of telomeres!

Her research was also perfectly timed, because methods for sequencing DNA were just being developed. Blackburn found that Tetrahymena's telomeres had an unusual nucleotide sequence: TTGGGG, repeated about 50 times per telomere.

Since then, scientists have discovered that the telomeres of almost all organisms have repeated sequences of DNA with lots of Ts and Gs. In human and mouse telomeres, for example, the repeated sequence is TTAGGG.

The number of telomere repeats varies enormously, not just from organism to organism but in different cells of the same organism and even within a single cell over time. Blackburn reasoned that the repeat number might vary if cells had an enzyme that added copies of the repeated sequence to the telomeres of some but not all chromosomes.

With her then-graduate student Carol Greider, now at Johns Hopkins University, Blackburn hunted for the enzyme. The team found it and Greider named it telomerase.

The telomerase enzyme, it turns out, consisted of a protein and an RNA component, which the enzyme uses as a template for copying the repeated DNA sequence.

What is the natural function of telomerase? As cells divide again and again, their telomeres get shorter.Most normal cells stop dividing when their telomeres wear down to a certain point, and eventually the cells die. Telomerase can counteract the shortening. By adding DNA to telomeres, telomerase rebuilds the telomere and resets the cell's molecular clock.

The discovery of telomerase triggered new ideas and literally thousands of new studies. Many researchers thought that the enzyme might play important roles in cancer and aging. Researchers were hoping to find ways to turn telomerase on so that cells would continue to divide (to grow extra cells for burn patients, for example), or off so that cells would stop dividing (to stop cancer, for instance).

So far, they have been unsuccessful. Although it is clear that telomerase and cellular aging are related, researchers do not know whether telomerase plays a role in the normal cellular aging process or in diseases like cancer.

Recently, however, Blackburn and a team of other scientists discovered that chronic stress and the perception that life is stressful affect telomere length and telomerase activity in the cells of healthy women. Blackburn and her coworkers are currently conducting a long-term, follow-up study to confirm these intriguing results.

THE OTHER HUMAN GENOME

Before you think everything's been said about DNA, there's one little thing we didn't mention: Some of the DNA in every cell is quite different from the DNA that we've been talking about up to this point. This special DNA isn't in chromosomes— it isn't even inside the cell's nucleus where all the chromosomes are!

So where is this special DNA? It's inside mitochondria, the organelles in our cells that produce the energy-rich molecule adenosine triphosphate, or ATP. Mendel knew nothing of mitochondria, since they weren't discovered until late in the 19th century. And it wasn't until the 1960s that researchers discovered the mitochondrial genome, which is circular like the genomes of bacteria.

In human cells, mitochondrial DNA makes up less than 1 percent of the total DNA in each of our cells. The mitochondrial genome is very small—containing only about three dozen genes. These encode a few of the proteins that are in the mitochondrion, plus a set of ribosomal RNAs used for synthesizing proteins for the organelle.

Mitochondria need many more proteins though, and most of these are encoded by genes in the nucleus. Thus, the energy-producing capabilities of human mitochondria—a vital part of any cell's everyday health—depend on coordinated teamwork among hundreds of genes in two cellular neighborhoods: the nucleus and the mitochondrion.

Mitochondrial DNA gets transcribed and the RNA is translated by enzymes that are very different from those that perform this job for genes in our chromosomes. Mitochondrial enzymes look and act much more like those from bacteria, which is not surprising because mitochondria are thought to have descended from free-living bacteria that were engulfed by another cell over a billion years ago.

Scientists have linked mitochondrial DNA defects with a wide range of age-related diseases including neurodegenerative disorders, some forms of heart disease, diabetes, and various cancers. It is still unclear, though, whether damaged mitochondria are a symptom or a cause of these health conditions.

Scientists have studied mitochondrial DNA for another reason: to understand the history of the human race. Unlike our chromosomal DNA, which we inherit from both parents, we get all of our mitochondrial DNA from our mothers.

Thus, it is possible to deduce who our maternal ancestors were by tracking the inheritance of mutations in mitochondrial DNA. For reasons that are still not well understood, mutations accumulate in mitochondrial DNA very slowly compared to chromosomal DNA. So, it's possible to trace your maternal ancestry way back beyond any relatives you may know by name—all the way back to "African Eve," the ancestor of us all!

Chapter 5

The Genetic System

ITS HISTORY DURING THE LIFE OF THE INDIVIDUAL AND IN THE PRODUCTION OF THE NEXT GENERATION

In earlier chapter it is described the discovery of the materials of heredity, and the methods by which it was proved that they do indeed affect the characteristics of organisms. These materials, it was shown, lie in the chromosomes. And it was seen that removing a chromosome, or substituting one for another, has great and varied effects on individuals.

The study of genetics is very largely therefore the study of these materials, and particularly the study of their physiological action, in reproduction and heredity. As a foundation for understanding their action, it is necessary to have clear ideas of the main features of their location and history, from the formation of the new individual, through its development to an adult, and thence to the formation of a new generation. This forms the subject of the present chapter. Though here heredity, variation and characteristics are not explicitly considered, it will be found that for their understanding, as taken up in later chapters, knowledge of the matters set forth in this chapter is essential.

The materials of heredity are gathered into a system of structures, which we may call the Genetic System. This system lies within the cells, and is as well defined in its constitu-tion and functions as is the nervous or muscular system. It includes the chromosomes; whether it includes also other parts lying outside the chromosomes will be considered later. For the present it is the chromosomes that will be followed through the life of the individuals, since it is known positively that these effect the development and characteristics of organisms. They will be traced, as material bodies, throughout the life of the individual, and into the next generation. Understanding of the action of the genetic system is largely dependent on having a clear picture of these matters.

To present this history clearly and systematically it will be best to set forth the main facts in a series of propositions, which will be further emphasized by giving to each a number or letter. We begin at the earliest condition of the young individual; that is, at the fertilized egg before it has divided.

THE GENETIC SYSTEM IN THE FIRST STAGE OF THE INDIVIDUAL; THE FERTILIZED EGG

1. As the new individual begins its separate life in the form of a fertilized egg, it contains, embedded in the cytoplasm of the cell, two nuclei, each with a set of chromosomes.
2. One of these nuclei is from the mother (the 'female pro- nucleus'), the other from the father (the 'male pronucleus').
3. The two pronuclei contain the same number of chromosomes (or that from the father contains one less; see the following). The number contained differs in different organisms; it will be convenient to call the number n. Thus each of the two pronuclei has a set of n (or n-i) chromosomes, and the entire egg has two sets, making zn (or in the male, 2w-i) chromosomes.

   The number «, present in each of the two pronuclei, is known as the haploid number. The number 2W (or 2«-i), present in the two together, is known as the diploid number. In different organisms the haploid number varies from i to more than 100.

4. In some of the fertilized eggs (those that will develop into males), in some organisms, the pronucleus from the mother contains one more chromosome than that from the father. This additional chromosome is known as X. In the eggs that are to produce females (in these same organisms) the two pronuclei contain the same number of chromosomes, each having an X.
5. In organisms in which the number n is the same for both pronuclei, frequently in the eggs that are to produce males one chromosome (X) in the maternal pronucleus is larger than the corresponding chromosome (Y) in the paternal pronucleus.
6. In some organisms the different chromosomes in each single pronucleus differ much in size and shape.
7. But the set of chromosomes in one of the two pronuclei is like the set in the other, as to size and form of the chromosomes, except in the case of the X and Y chromosomes above mentioned.
8. The two pronuclei are at first separate; they approach one another, come in contact, their membranes dissolve, and the two groups of chromosomes come closer together, forming a single group of 2« chromosomes (or 2«-i). This is the diploid group.
9. Later this group becomes surrounded by a single membrane, constituting thus the single nucleus of the fertilized egg, containing the diploid number of chromosomes.
10. Thus in the single nucleus of the fertilized egg, half of the chromosomes are from the mother, half (or one less) from the father.
11. Sooner or later the corresponding chromosomes of the paternal and maternal sets mate or conjugate, forming a pair. In some animals, as in the flies, this pairing of the chromosomes takes place soon after the union of the two pronuclei (in the 2-cell stage).

12. Thus the nucleus of the fertilized egg contains n pairs of chromosomes, one member of each pair paternal in origin, the other maternal.
13. In some organisms (the flies and others), the two chromosomes of a pair (maternal and paternal) remain paired throughout the life of the individual, and in all his cells.
14. In other species the two members do not pair till later (when germ cells are formed). But finally they mate, in practically all organisms.

    We shall later examine what happens in mating of the chromosomes.
15. Even in species in which the maternal and paternal members are not at all times side by side, it is often possible to determine under the microscope which two chromosomes form a pair, since the pairs are of different sizes and forms, while the two members of the same pair are of the same size and form.

    The paired condition of the chromosomes turns out to be of great physiological importance.

THE CHROMOSOMES IN THE CELL DIVISIONS THAT PRODUCE THE BODY

1. The egg divides into 2 cells, these into 4, and so on, to 8, 16, and the like, till a great number of cells are formed, making up the body of the adult organism.
2. At each cell division, each of the 2« chromosomes divides into 2, by splitting lengthwise.
3. One half of each chromosome passes into each of the two cells formed by division.
4. There, each half grows to form a complete chromosome. Thus the chromosomes reproduce by fission, like Protozoa.
5. Consequently each cell formed contains the full set of 2« (or 2«-i) chromosomes, half of them maternal, half paternal.

Ultimately therefore every cell of the body contains the set of 2w (or 2«-i) chromosomes, half maternal, half paternal, in origin. (There are some exceptions to this statement.)

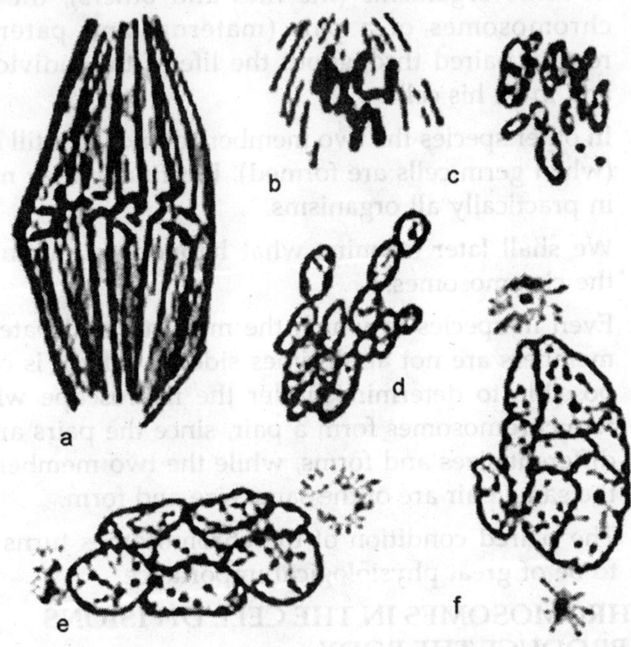

Fig. Transformation of the condensed chromosomes (a) by absorption of fluid into vesicles which slowly enlarge (b, c, d, e,f), finally constituting the nucleus. The Fig show the process in the cells of the egg of a fish, Fundulus.

THE CHROMOSOMES BETWEEN THE CELL DIVISIONS, WHEN THE CELLS ARE NOT DIVIDING

1. At the time of cell division the chromosomes are in the condensed condition.
2. After completion of cell division, each chromosome absorbs fluid from the cytoplasm, grows, enlarges, and becomes a small vesicle.

   This is a slow, gradual process, that can be seen under the microscope.

3. The vesicles become so large that their boundaries touch, and are pressed together, the boundaries becoming very indistinct or completely disappearing.
4. The entire set of zn (or 2fl-i) vesicles derived from the chromosomes constitutes the nucleus (what is called the 'resting nucleus').
5. In some organisms the entire set of zn chromosome vesicles can be dimly seen and counted, even in this resting stage. In others this cannot be done.
6. As any cell gets ready to divide again, a new, small chromosome is laid down in each of the vesicles formed by the previous chromosomes.
7. When these new chromosomes first appear, they are usually irregular granules, which gather into a thread, having thickenings at intervals. In this condition they are called the skein or spireme.
8. These threads gather together and condense into the chromosomes of the type seen at cell division.
9. The remainder of each large chromosomal vesicle dissolves in the cytoplasm of the cell. Here it doubtless produces chemical changes of importance.

   Thus after every cell division the chromosome takes up a quantity of fluid from the cytoplasm, and later gives it off again, doubtless in modified form. This is probably one of the most important processes in development. It is presumably in this way that the chromosomes influence the development and the characteristics of the organism. They take up parts of the cytoplasm, change it, and give it off again. In this way are probably made the different kinds of tissues: muscles, nerves, bones, and the like. The chromosomes are thus continually active in the process of manufacturing the body (a matter to which we return later).
10. While thus in vesicular form, and active in the physiological processes of the cell, the chromosomes

as such have in most cases disappeared; they are no longer visible (though in exceptional cases they can still be detected). When later they reappear they show definite relations to the chromosomes that were visible before disappearance, in the following particulars:

a. The same number reappears. If the number is changed experimentally, as has often been done, it is the altered number that reappears.
b. The chromosomes reappear in the same set of diverse forms and sizes as before. This is very striking in organisms like Drosophila in which the different pairs of chromosomes differ much. In hybrids, often the maternal and paternal sets differ greatly in size and form; each set reappears in its own type.
c. The chromosomes reappear in the original grouping and arrangement. When the chromosomes that disappeared were in pairs, as in Drosophila they reappear paired as before. Often at disappearance the maternal and paternal sets are in two separate groups they reappear in these same groups. Often two particular chromosomes that are close together or intertwined reappear in this same condition.

Thus, even in cases in which the chromosomal vesicles are not separately detectable during the 'resting stage', it is obvious that each chromosome reproduces itself, in the same form, size, and position.

n. All this is repeated at every cell division (though the arrangement or relative position of the chromosomes gradually changes in the course of many cell divisions). Thus finally each cell of the body contains the complete set of 2/2 chromosomes, half from the mother, half from the father.

THE PRODUCTION OF GERM CELLS, IN RELATION TO THE MATERIALS OF HEREDITY

Having followed the main features in the behaviour of the chromosomes in forming the body of the individual, we next examine what happens when the individual reproduces:

the processes in forming germ cells and in producing offspring. In dealing with heredity, it is particularly important to have these correctly in mind.

1. The individual begins as a single cell, the fertilized egg.
2. It divides into many cells, some of which produce the body, while others produce the germ cells (G).
3. At an early stage in development there is usually among the numerous cells one that is later to produce all the germ cells; this is the primordial germ cell.
4. The primordial germ cell in most organisms has the same set of chromosomes and genes as have the body cells. (There are exceptions to this in certain organisms.) That is, the primordial germ cell has n pairs of chromosomes, making 2« (or 2fl-i) chromosomes. Of these, n are maternal in origin, while n (or n-i) are paternal in origin.

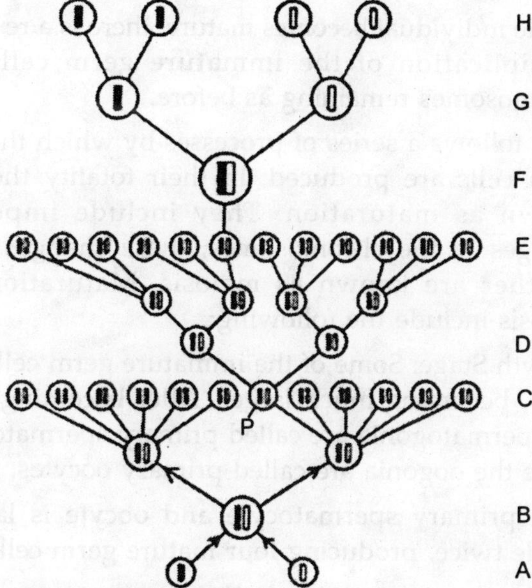

Fig. Diagram to illustrate the relation of body cells and germ cells) and the origin of both from the fertilized egg. Only a single pair of chromosomes is represented in each cell. At A are the two gametes (sperm and ovum) which unite to produce the fertilized egg B; each gamete contains one chromosome of the pair. The fertilized egg B divides, producing body cells C, among which is the primordial germ cell P. The primordial germ cell divides to produce many immature germ cells (E). The immature germ cell enlarges, the two chromosomes of a pair conjugating (F), then divides into four germ cells (G, H), each containing but one chromosome of the pair. (In the female three of the four germ cells (H) thus formed are not functional.}

5. The primordial germ cell divides many times, to produce many immature germ cells. These still contain the full two sets, n pairs, of chromosomes.

   The immature germ cells in the male are the spermatogonia, while those in the female are oogonia.

6. There is usually a period of youth of the organism, in which the immature germ cells remain at rest, not dividing actively.

7. As the individual becomes mature, there is a renewed multiplication of the immature germ cells, the chromosomes remaining as before.

   Next follows a series of processes by which the final germ cells are produced: in their totality they are known as maturation. They include important changes in the chromosomes; these changes taken together are known as meiosis. Maturation and meiosis include the following:

8. Growth Stage: Some of the immature germ cells now grow, becoming much larger. After becoming large, the spermatogonia are called primary spermatocytes, while the oogonia are called primary oocytes.

9. Each primary spermatocyte and oocyte is later to divide twice, producing four mature germ cells.

   While the growth and division are occurring, important processes occur in the chromosomes, as follows:

10. The chromosomes pair if they have not already done so; that is, the corresponding paternal and maternal chromosomes place themselves side by side close together.
11. While pairing, the chromosomes are in the form of very long slender threads, which have thickenings at intervals. There are corresponding thickenings side by side in the two threads.
12. The two chromosomal threads (maternal and paternal) forming a pair become intimately united. This process is known as the conjugation, or synapsis, of the chromosomes. (During conjugation the two chromosomes may exchange parts, by 'crossing-over, as will be described later.)

    This intimate union of the maternal and paternal chromosomes appears to be the ultimate act in the union of the sexes.
13. Next occur certain processes, the general result of which is simple, although the processes are complex. The final result is that which would be produced if the immature germ cell now divided but once into two final germ cells and, in so doing, the two chromosomes of each pair separated, one going to one of the two germ cells, one to the other. Thus each mature germ cell contains finally one chromosome (the maternal one or the paternal one) from each pair. The total number of chromosomes in the germ cell is now one-half the number originally present in the immature germ cell; it is n in place of 2». This is the process known as reduction of the number of chromosomes.

    The actual processes are more complex, and are somewhat confusing. In following them, the final result just set forth should be kept in mind. They are as follows:
14. During conjugation the two long chromosomal threads shorten and thicken.

15. And while this is occurring, each of the two splits into two, so that there are now four partly united threads, two maternal, two paternal, in origin.
16. The shortening and thickening continues till the four threads have gathered into four closely united lobes. These are known as 'groups of four', or 'tetrads'.

    Since each pair of chromosomes becomes one tetrad, there are n tetrads. In each tetrad, two lobes represent a maternal chromosome, while two represent a paternal chromosome.

    Some males have one chromosome (X) without a mate; this does not conjugate, but otherwise goes through the same process as the others, forming a two-lobed instead of a four- lobed body.
17. Now the large cell containing the n tetrads divides quickly, twice in succession ('maturation divisions'), thus producing four final germ cells (sperms or ova)
18. In these two maturation divisions, each tetrad divides twice, in such a way that the four lobes separate into the four different germ cells formed.
19. Thus two of the four germ cells receive from each tetrad a maternal lobe only, while two receive a paternal lobe only.
20. These lobes transform each into a chromosome like that from which it was derived.
21. Thus each final germ receives either a maternal or a paternal chromosome from each of the n pairs that were present in the immature germ cells. Each therefore contains but n single chromosomes, in place of the n pairs (or zn single chromosomes) that were present in the parental cells. This is the process of reduction of the number of chromosomes, from 2n to n.

    As set forth in paragraph 13 above, the net result is the same as would be produced by the division of each immature germ cell into two instead of four, one chromosome from each pair passing into each of the two germ cells produced. For many practical purposes, one can think of germ-cell formation as occurring in this way.

22. In the female, all the germ cells produced have the same number of chromosomes (fl), since each receives an X chromosome. In some species, in which the male has an odd number (2«-i) of chromosomes, half the germ cells receive n chromosomes, the other half receiving n-i.
23. In the male, all the four germ cells produced by the two maturation divisions are functional; they transform into gametes, known in the males as sperms.
24. In the female, three of the four cells formed by the two maturation divisions are small and without function; they are known as the polar bodies. The fourth is large, and is the functional gamete or ovum, which may be fertilized and may develop into a new individual.
25. A sperm with n (or n-i) chromosomes unites with an ovum containing n chromosomes, giving again a fertilized egg, or zygote, containing n pairs of chromosomes.
26. Different germ cells from the same parent, and different fertilized eggs produced by the same two parents, receive different combinations of chromosomes. This results from the following facts:
    a. Each parent carries n pairs of chromosomes, one chromosome of each pair being maternal, the other paternal, in origin.
    b. As we saw in the preceding chapter, the paternal and maternal chromosomes of a pair often differ in their effects. We may therefore designate them differently, calling one of them A, the other a, so that the pair is Aa.

c. The chromosomes of the different pairs are often diverse in size and form; and, as will be shown later, they are also diverse in their functions. Therefore we may designate the chromosomes of the different pairs by different letters. Thus the maternal chromosomes of the different pairs may be represented by the capital letters A, B, C, and so on, while the paternal chromosomes of the same pairs may be represented by the small letters a, b, c, and so on. The series of pairs in the cells of the parent are thus Aa, Bb, Cc, and so on, for n pairs.

d. As before seen, each final germ cell receives one chromosome from each pair. It may receive A or a from the pair Aa; B or b from the pair Bb, and so on.

e. The different pairs are independent in their distribution to the germ cells. Thus from the two pairs Aa and Bb a given germ cell may receive any of the four combinations AB, Ab, aB, and ab; and similarly for other pairs. Any one combination occurs as frequently as any other.

f. Thus with relation to one pair, as Aa, two types of germ cells are possible, A and a; similarly two types are possible from the pair Bb. Either type from Aa may be combined with either type from Bb, giving four diverse types from the two together, as illustrated above. Each additional pair multiplies the number of possible combinations by 2.

g. Thus if the parent has n pairs of chromosomes, and the chromosomes of maternal origin differ in effect from those of paternal origin, the number of different types of germ cells producible as a result of this situation is 2^n, these having diverse combinations of chromosomes in the different germ cells produced by that parent. Thus in Drosophila melanogaster, which has four pairs of chromosomes, the number of diverse types of germ cells that may be thus formed is 2, or 16. If the four pairs are Aa, Bb, Cc, and Dd, then in the different germ cells there are such combinations as ABCD, AbCD, aBcd, aBcD, and so on. An organism with 10 pairs of chromosomes similarly produces 1024 diverse types of germ cells. In man, having 24 pairs of chromosomes, the number of diverse types producible is 2, or more than a million.

(As will be shown later, the number of diverse types of germ cells producible is greatly increased beyond these Figs by the fact that the two chromosomes of a pair may during conjugation exchange parts in various proportions.

27. Thus any single individual produces many diverse types of germ cells, containing diverse combinations of its chromosomes. This is true both for the sperms and the ova.
28. In fertilization, sperms and ova having the various diverse combinations of chromosomes unite at random. Thus the many fertilized eggs, and the offspring developed from them, bear different combinations of chromosomes. In consequence, as will be shown later, they may develop differently, giving many diverse types of individuals. The numbers of different types thus producible by g, given pair of parents is dealt with in later chapters.

CERTAIN TERMS DEFINED

In connection with the above account of the history of the materials of heredity, it will be well to observe the meaning of certain common terms, whose use is very convenient; they will be employed in later pages.

Gametes: The mature germ cells (ova and sperm) are known as gametes. In the gametes the chromosomes are single instead of in pairs: that is, the gametes contain but n chromosomes, in place of zn. Cells having thus but one set of chromosomes are said to be haploid, while the usual cells, with chromosomes in pairs, are diploid.

The entire process of producing gametes is known as gametogenesis. The later processes, by which the four mature germ cells are produced, are spoken of as maturation. The chromosomal changes during maturation, resulting in reduction in the number of chromosomes from 2n to n_y are known as meiosis.

Zygote: The fertilized egg, formed by the union of two gametes, is the zygote. This term is often applied to the individuals developed from the zygote. The zygote is diploid: its chromosomes are in pairs, their total number being 2fl. The union of the gametes to form the zygote is spoken of as the fertilization of the ovum.

Heterogametic: The individuals of one sex (the male in the types described above) form two kinds of gametes, one containing an X-chromosome, the other not. Such individuals are said to be heterogametic (or sometimes the term digametic is employed in the same sense).

Autosomes and Sex Chromosomes: The chromosomes X and Y, since they play a prominent part in the production of sex, are known as sex chromosomes. All the other chromosomes are known as autosomes. Thus Drosophila melanogaster has one pair of sex chromosomes and three pairs of autosomes.

Chapter 6

Operation of the Genetic System, as Exemplified by its Relation to Sex

Development and characteristics are affected in innumerable ways by the materials of heredity, as is shown by the results of substituting one chromosome for another. Among the most striking alterations so produced are changes of sex. A study of the relations of chromosomes to sex furnishes an excellent introduction to the physiology of the genetic system: it indeed leads far into the subject. The relations of chromosomes to sex will therefore be taken up first, and treated with some fullness, as revealing the nature of chromosomal action.

It was set forth that in many organisms the sex of the individual to be produced by an egg may be changed by altering the chromosomes which the egg contains. We wish to inquire how the chromosomes operate in producing sex differences.

TWO GROUPS OF ORGANISMS, HAVING DIFFERENT RELATIONS OF CHROMOSOMES TO SEX

In examining the question proposed in the preceding paragraph, we come upon the striking fact that in different groups of organisms there are very different relations of chromosomes to sex. There are two main groups, showing contrasted relations in this matter. One group, which includes the larger number of animals and plants, In these, the males have an unequal pair of chromosomes, X and Y (or they may lack completely the smaller chromosome Y), while in the females the corresponding pair consists of two equal chromosomes, XX. After this group had become known, the remarkable fact was discovered that there is another group of organisms, including the birds and some moths, in which this situation is reversed. In this second group it is the females that have the unequal pair of chromosomes, while the males have the corresponding equal pair. This is obviously a fact requiring consideration in all attempts to understand how the chromosomes operate in producing diversity of sex. The two groups may be characterized as follows:

Group I The males contain one pair of dissimilar chromosomes, X and Y; in some cases Y is lacking entirely. In addition they have a number of pairs, the autosomes, in which the two members of each pair are alike. The females carry an equal pair XX in place of the unequal (X +Y) pair of the males; they also have, of course, the usual autosomes. The conditions in Group I are represented in Fig.

Thus in Group I the males are heterogametic, producing two kinds of sperms in equal numbers, one set of sperms bearing *n* autosomes plus X, the other bearing *n* autosomes without Xýÿ-but in many cases with a Y. In the females all the gametes (ova) are alike as to their chromosomesýÿeach containing *n* autosomes plus X. Sex is determined by the type of sperm that enters an ovum. A sperm of the constitution (autosomes + X) uniting with an ovum (autosomes + X) gives a female (2 sets of autosomes + XX). A sperm of the constitution (autosomes 4- Y) uniting with an ovum (autosomes + X) gives a male (2 sets of autosomes + XY). (In all such cases the Y may be lacking.)

To Group I belong many groups of organisms, including man. The following groups of animals are among those that show the relations of Group I: echinoderms, nematodes, molluscs, most insects, arachnids, myriapods, fish, mammals.

Four main subtypes may be distinguished in Group I, differing in the condition found in the males. These are:

Subtype i. Male has X but no Y.

Subtype 2. Male has a large X and a small Y. To this group belongs man. He has 23 pairs of autosomes, an X, and a small Y.

Subtype 5. The male has an X, and a Y that is of different form and function from X .

Subtype 4. The Y chromosome does not differ in form or size from X, but experiments of the kind to be described later show that it differs from X in function.

Other conditions, not classifiable under any of these four subtypes, occur in some organisms.

Group II.The females have a pair of dissimilar chromosomes, in addition to the usual pairs of autosomes, while the males have a corresponding pair of similar chromosomes, in addition to the autosomes. The dissimilar chromosomes of the female are often called X and Y, as in the other group, while the similar ones of the male are called XX. But, by some authors, the dissimilar pair in this group are designated ZW, while the corresponding similar pair are ZZ.

In Group II, the females are heterogametic, producing two types of ova (autosomes+ Z and autosomes + W). In the males, all the sperms are alike, with the constitution (autosomes+Z). The sex is determined by the type of ovum, not by the type of sperm, that enters into the fertilized egg. A sperm (autosomes + Z) uniting with an ovum (autosomes + Z) gives a male (2 autosome sets -f ZZ). The same type of sperm (autosomes-f Z) uniting with an ovum (autosomes + W) yields a female (2 autosome sets + ZW).

To this group belong birds and certain moths. The relations of the chromosomes to sex have been less fully studied in this group than in Group I.

PHYSIOLOGY OF SEX DETERMINATION IN GROUP I

The conditions in Group I have been extensively studied; they throw much light on the nature of the physiological action of the chromosomes, in their effects on development.

In Group I, as before seen, the males and females both contain in their cells two sets of autosomes (one set derived from each parent), and also an X-chromosome. They differ in the fact that the female contains an additional X, while the male has in place of this a Y; or in some species the Y is lacking entirely.

What part do these combinations of chromosomes play in producing the differences of sex? The way to determine this is to alter the chromosome combinations experimentally, and observe what differences this makes. This has been done extensively, originally by G. B. Bridges in the fruit-fly, Drosophila melanogaster. The results are of great interest; they will therefore be examined.

ALTERATION OF THE CHROMOSOME COMBINATIONS IN THE FRUIT-FLY, AND ITS EFFECT ON SEX

It is important in following this work to have clearly in mind the chromosomal conditions in the organism used as an example. In this species of fruit-fly there are four pairs of chromosomes, three of them being autosomes. In addition, the female has two large straight X's, while the male has one straight X and a hooked Y, of about the same size as the X.

The autosomes as well as X and Y are important for sex, as it turns out. It will be convenient to indicate the three autosomes simply by the letter A. Thus the male produces two types of sperms, AX and AY, while the ova produced by the female are all alike, having the constitution AX. When the ovum AX is fertilized by a sperm AX, the result is AAXX, a female. But when the ovum AX is fertilized by the sperm AY, the result is AAXY, a male.

THE EFFECT OF CHANGING THE NUMBERS OF X AND Y CHROMOSOMES PRESENT (NON-DISJUNCTION)

Bridges discovered that sometimes, in forming the ova, an accident occurs of such a sort that the two X's do not separate into different ova, but both go together into one ovum, leaving another ovum without an X. Thus two kinds of ova are produced, AXX and AO (using O to signify the absence of X or Y).

This failure of the two X's to separate is called non-disjunction of the X-chromosomes. Later a race of fruit-flies was discovered by L. V. Morgan in which the X-chromosomes are stuck together, so that they almost never separate. In this race germ cells are practically always formed in the way just described. So it has been possible to study the results thoroughly, since thousands of offspring have been obtained. These results are of great importance and interest.

From such individuals we have then two kinds of ova, one AXX, with two X's; one AO, with no X. What will happen when these unite with the usual two kinds of sperm, AX and AY?

When the ovum AXX unites with a sperm AY, we get offspring AAXXY containing two X's and also a Y. Will such an individual be a male or a female? The results show that it is a female.

This result shows at once certain important things about sex determination,

1. The presence of Y (found normally in males) is not sufficient to produce a male.
2. It is the presence of two X's, not the absence of Y, that yields a female.
3. The same sperm AY that would normally produce a male, yields a female if the ovum contains two X's.

When the ovum AO, without X, unites with a sperm containing X (that is, AX), we obtain a fertilized egg AAXO that has but one X, and no Y. What sort of an individual will this be?

Observation shows that this combination produces a male, although it has no Y. Certain other important points are thereby demonstrated. (4) The Y-chromosome is not required in order to produce a male. (5) It is the lack of one X that gives origin to a male. (6) The sperm that contains X, which normally produces a female, yields a male if it enters an ovum containing no X. (7) When the usual autosomes are present two X's give a female, one X a male, whether the Y-chromosome is or is not present in addition.

Certain other combinations are produced. When an ovum AO, lacking X, unites with a sperm AY, containing Y but no X, the fertilized eggs AAY do not develop. () This shows that the presence of one X is necessary for development.

When an ovum AXX, with two X-chromosomes, unites with a sperm AX that also contains an X, the resulting individual AAXXX contains three X's, but only two sets of autosomes. The results are as follows: Most such individuals with XXX do not develop. A few develop, but are abnormal; development is disturbed by the presence of the additional X. The abnormality consists in an accentuation of the structural peculiarities that distinguish the females from the males. The individuals seem as it were to be more female than those containing but two X's. Bridges calls them super-females.

From these experiments we discover that when two normal sets of autosomes are present, lack of an X is fatal: the presence of one X causes the production of a male; two X's cause the production of a female; three X's give rise to an abnormal 'super-female'.

One further fact should be mentioned. The males containing a single X but no Y, while well formed, are sterile; they do not produce functional sperm. Thus the chromosome Y is not entirely without function; it is necessary for the production of normal male germ cells.

Still other combinations of chromosomes can be produced by mating together in various ways the different types of individuals just described. In the following list are given the various combinations that have been produced, with the nature of the individuals that they yield.

AAXY, normal male.

AAXX, normal female.

AAXXY, typical female.

AAXXYY, typical female.

AAXO, male, sterile.

AAYO, does not develop.

AAXXX, abnormal, 'super-female.

All of these results show that changing the number of X's present is what changes the sex, while changing the number of Y's does not change the sex.

In all these cases there are present two sets of autosomes, AA, in addition to the X's and Y's. Have these autosomes any effect on sex? What would happen if the numbers of autosomes were changed? To this question we now turn.

THE EFFECT OF CHANGING THE NUMBER OF SETS OF AUTOSOMES PRESENT

Bridges discovered further that sometimes the number of autosomes becomes changed, and he determined the results of such changes. In the normal individuals there are two sets of autosomes, one from each parent; that is, the chromosomes are in pairs, AAXX. Certain individuals in the fruitfly were discovered to have their chromosomes in sets of three instead of in pairs. Such individuals are known as triploids; their chromosomes may be represented as AAAXXX. How these individuals were produced is not known. It seems probable that in some way germ cells were produced in which the two chromosomes of the pairs did not separate, so that these germ cells were AAXX. These being fertilized by the usual sperm, AX, gave the triploids. It is important to understand that the existence of these triploids is determined under the microscope. Their chromosomes in microscopic preparations are seen to be in threes.

When such triploids form germ cells, some of the latter are of the usual haploid type, AX, while others have their chromosomes in pairs, AAXX; that is, they are diploid. Some of both types of course have Y-chromosomes in place of the X's, so that there are germ cells AY, AX, AAYY, AAXY, and AAXX. Such germ cells when they unite produce many different chromosome combinations, some with autosomes in pairs, others with autosomes in threes or even in fours; and with different numbers of X's and Y's. Many individuals having such combinations have been produced and studied. This has given opportunity to determine the effect of varying the numbers of sets of autosomes, as well as of varying the X's and Y's.

The more important different combinations, with the nature of the individuals produced, are given in the following list:

Autosomes Sex chromosomes Result

2 sets, AA XX Normal females

3 sets, AAA XXX Normal females 4sets, AAA XXXX - Normal females 3 sets, AAA XX 'Intersex'

3 sets, AAA XY 'Super-males'
2 sets, AA XXX 'Super-females'

Here we find that while two X's normally give a female, if they are combined with three sets of autosomes they do not give a female, but an individual intermediate between the male and female; an 'intersex'. It is clear therefore that the autosomes influence sex, as well as do the X's. Furthermore, it is not the absolute number of autosomes present that gives a particular effect, but the number of sets of autosomes in relation to the number of X's. Three sets of autosomes with but two XX's gives an intersex, while three sets of autosomes with three X's yields again a normal female. The sex produced depends upon a balance between the autosomes and the X's. When the number of sets of autosomes equals the number of X's (whether two, three or four) a female is produced. When the number of sets of autosomes exceeds the number of X's, there is a tendency for males to be produced. Two sets of autosomes plus one X gives the normal male. Three sets of autosomes plus two X's yields a combination in which the proportional excess of autosomes is not sufficient to produce complete maleness; it yields intersexes. But when three sets of autosomes are present with but one X, the scale is tipped too far in the direction of maleness. Abnormal individuals are produced in which the characteristics that distinguish males from females are accentuated beyond the normal; Bridges calls these 'supermales'. And similarly, if three X's are present with but two sets of autosomes, the balance is tipped in the opposite direction, and 'superfemales' are produced.

We may summarize the special results of these observations as follows:

1. Both autosomes and X-chromosomes take part in determining the sex. Altering either class of chromosomes changes the sex.
2. Which sex is produced depends upon a balance between the number of sets of autosomes and the number of X's.
3. A preponderance of the number of X's tends to produce females; a preponderance of the number of sets of autosomes tends to produce males.
4. By altering the balance in ways not commonly occurring, various intermediate and extreme conditions as to sex are produced.

METHOD OF OPERATION OF THE CHROMOSOMES IN PRODUCING SEX DIFFERENCES

The results just set forth are of great significance for the understanding of the method of operation of the chromosomes in development. Their bearing will best be appreciated by bringing out certain other relations that are found in the observations above described, on the relations of chromosomes to sex.

1. As before seen, changing the number of X's, when the autosomes are left unchanged, changes the sex.Two XX's produce a female, one X a male.
2. Thus the female XX contains everything that is necessary for producing a male. It differs in having

'two doses' of X in place of one. By removal at an early stage of development of one of these 'doses', it should be possible to transform the female into a male.

3. The male, XY or XO, contains all the kinds of materials required to produce a female. It contains X, as does the female; but it has only 'one dose' of X in place of two.

   It should be possible therefore to convert the male into a female if the single X-chromosome of the fertilized egg could be caused to divide into two, while the other chromosomes remained undivided, thus giving an individual with chromosomes AAXX.

4. Thus males and females do not differ at the beginning by containing different kinds of materials. Both demonstrably contain the same kinds of materials. The difference lies in the fact that the female has two centres of growth and multiplication for a certain material (X), while the male has only one centre of growth for that same material. Such a difference is commonly spoken of as a difference in 'balance' among the chromosomes.

If this material (X) has a certain type of effect, the difference between a condition with two centres of action for that material, and a condition with but one, might well make a great difference to the chemical and physiological processes occurring. In every cell, throughout life, the processes in the female are such as result from the interaction of these two centres with the other cell contents, while in every cell of the male the result is that produced by the interaction of but one such centre with the other contents.

The result of this difference in balance of the chromosomes is in fact to produce the very great differences, structural, physiological, mental, that distinguish the female from the male. The fact that so many and so great differences, of so many diverse kinds, are brought about without any original difference in the kinds of materials present, but only in balance, only in the number of centres of growth for certain materials, is of the greatest significance for the nature of the processes of development; it should never be lost sight of in considering the facts of heredity. Other differences produced by diversity of chromosomes, to be taken up later, may perhaps be brought about in a similar way.

THROUGH WHAT MEANS DO THE CHROMOSOMES PRODUCE SEX DIFFERENCES?

How can differences between chromosomes produce such great differences as those shown by individuals of different sexes? What is their method of action in bringing about sex differences?

Although a complete answer to these questions cannot be given, much is known that bears on them. Many things have been discovered as to the means through which chromosome differences produce sex differences, and these things are most instructive for forming a conception of the nature of development and heredity.

ACTIVITY OF THE CHROMOSOMES IN DEVELOPMENT

As before seen, the chromosomes are very active bodies. They are continuously working at the cytoplasm of the cells, interacting with it, changing it. In the very young cells, just after cell division, the chromosomes are small compact bodies. They begin to absorb material from the cytoplasm around them. They thus enlarge, becoming vesicles filled with substance. Each becomes doubtless a hundred times as large as before. In time the vesicles become so large that they are crowded together to form the large nucleus.

What becomes of this large amount of fluid taken in by the chromosomes? It undoubtedly becomes chemically changed by interaction with the chromosomes. There is evidence of this in the chemical reactions of the material of the nucleus as compared with those of the cytoplasm, and in certain other processes that will be discussed later.

The nucleus thus has filled itself with cytoplasm, which has been acted on, changed, by the chromosomes. Then, before the next cell division, it pours this altered material back into the cytoplasm. The outside membrane of the nucleus dissolves and all the materials pass out, except that there remains still a small concentrated chromosome, from each one of the former vesicles.

This taking in of cytoplasm, altering it, and giving it off again, is repeated at every cell division, so that it is continuously going on throughout development; it is occurring in every cell of the body. In this way the cytoplasm becomes gradually changed. Different kinds of materials are produced in the different cells; the different tissues and organs of the body are thus ultimately formed.

Thus it might be expected that sex differences would be produced through the fact that different chemical reactions occur when the cells contain but one X, from those that take place when two X's are at work. And there is proof that this is true. To illustrate these matters, we may summarize what happens in the development of sex differences in mammals.

1. In mammals the two sexes differ at the beginning of their lives, in the fertilized egg, as to the number of X-chromosomes they contain. Both contain the usual two sets of autosomes; in addition the female has two X's, the male but one X. What difference in development does this make?
2. At first the two sexes develop alike, so far as can be seen, so that the individuals are in an 'indifferent' condition, with respect to sex.

   In this indifferent condition, the embryos develop so far that head, limbs and all systems of organs are distinguishable: in the rabbit this condition is said to continue for about 14 days.
3. While in this indifferent condition, certain cells are set off as germ cells. These are gathered into a strip

of small cells called the germ gland, lying on the dorsal surface of the body wall. Some of the small cells become much larger than the others; these are called genital cells. The genital cells are to produce later the new germ cells (sperm or ova) of this individual. They are alike in the two sexes (except as to their X-chromosomes of course).

4. Now the germ glands in individuals of which the cells contain but one X in the males begin to develop in a distinctive fashion: the cells arrange themselves in 'cords' and take on a characteristic appearance. At the same time the germ glands of individuals whose cells carry two X's the females remain in the indifferent condition. Later, in the males the genital cells divide into small cells, which become the mother cells of the sperms; while in the females they enlarge, later to produce ova. These differences in development are due directly to the fact that one set of cells contains a single X, the other set two X's.

   As seen above, there is a period in which the male has developed distinctive characteristics, while the female is still in the indifferent condition. This turns out to be a matter of importance, as will be brought out later.

5. The two sexes continue to develop differently to adult life. The male germ gland becomes the testis, which produces sperms; and the body of the male develops differently from that of the female, producing the male secondary sex characters (mane, beard, greater size and the like). The female germ gland becomes the ovary, which produces ova; and the body develops the female secondary sex characters (mammary glands, diverse body form and size and the like).

   Thus the male and female have become very diverse, differing in a great number of ways, both structurally and physiologically.

All these later differences result in some way from the original difference in the X-chromosomes: from the presence of but one X in one set of individuals, of two X's in the other. Our present question is: through what means does the chromosome difference produce the later differences?

The answer to this question is approached through a series of experiments, the results of which we may summarize:

Experiment

Remove the germ gland (testis or ovary) from the very young individual, before the later sex differences are produced.

The result is that the later sex differences do not appear. Both sexes remain nearly or quite in the indifferent condition, so far as sex differences are concerned, though they continue to grow and develop in other ways.

7. It follows that the later sex differences are produced through the action of the germ glands (ovary or testis). The body cells of the two sexes still retain the differences as to the number of Xchromosomes present, but without the germ glands this does not result in the production of the later sex differences.

   How do testis and ovary act in producing the later sex differences? This is tested by another experiment:

8. Experiment.Interchange the germ glands of the two sexes. Remove the ovary from a young female, and replace it by transplanting to the female body a testis taken from a young male.

   Similarly, remove the testis from a young male and replace it by an ovary taken from a young female.

   These experiments are difficult, but have been successfully carried out in rats by a number of investigators.

   Result.The female body (XX) containing a testis (XO) develops the male secondary sex characters! The male body (XO) containing an ovary (XX) develops the female secondary sex characters (such a male develops a mammary gland that may produce milk; it may suckle the young).

9. The results of the experiments show that it is the nature of the germ gland (whether of the XX or the XO type) that determines (mainly or entirely) what later sex characters shall be produced. A body whose cells contain but one X will produce either male or female secondary sex characters, depending on whether it contains a germ gland having cells of the constitution XO or cells of the constitution XX. And the same is true for the body whose cells bear two X's.

10. By what means do the germ glands affect the development of the body?

    The germ gland is small and located at a definite place in the body. The secondary sex characters appear in many parts of the body, often situated far from the germ gland. Thus the germ gland influences parts at a distance from itself. How is this done?

11. The suggestion that comes most readily to mind is that each type of germ gland produces a characteristic secretion, which passes throughout the body and influences the develop ment of all parts. The secretion from the testis (XO) would thus produce one type of development (toward the male), the secretion from the ovary (XX) another type. Such suggestions have often been made.

12. Confirmation of the suggestion of diverse secretions circulating in the body: the Treemartin'.When twins of opposite sexes occur in cattle, the male twin is a normal, welldeveloped male. But the female twin is usually an 'intersex'; it is intermediate in structure between the male and female. It has many male characteristics, together with some female characteristics. There are different grades of intermixture, in different cases, and sometimes the female is normal.

What is the cause of this partial transformation of the female into a male?

It was found by F. R. Lillie that in all such cases there is an intercommunication of the blood system of the two twins, so that blood flows from the body of the male twin into the body of the female, and vice versa. But, in cases where the female is normal, there is no intercommunication of the blood systems.

From this it appears clear that the cause of the partial transformation of the female into a male is this intercommunication. Blood flows from the body of the male into that of the female, causing the latter to develop in the direction of maleness.

But why should it change the female only, not the male? Why does the latter not become intermediate also?

The answer to this question is given, with great probability, by the fact noticed on an earlier page, that the male begins the development of its distinctive sexual characteristics earlier than does the female. The individual is recognizable as male, before the female changes from the juvenile condition. Hence the male begins to produce its characteristic secretion first. This then circulates into the female body, and transforms it in the direction of maleness. As a result the female never produces its characteristic secretion, and so the male is not influenced.

Such secretions circulating in the blood, and influencing development, are known as hormones.

Can such male and female hormones be extracted from the blood, and used experimentally to produce male or female characteristics? This has been done to a certain extent. The hormones have not yet been obtained in a pure condition. But enough has been done to show that male and female hormones do exist, that they circulate in the body, and that they affect the sex characteristics.

(The account given above relates to mammals. In birds also sex hormones have been discovered, playing a role similar to that of the sex hormones in mammals, though with some striking differences. In insects, on the other hand, hormones do not appear to play a part in producing sex differences.)

THE QUESTION ANSWERED: THE MEANS BY WHICH THE CHROMOSOMES AFFECT SEX

The facts just summarized as to sex development in mammals provide a general answer to the question as to the method of operation of the chromosomes in determining sex. The answer is that the chromosomes determine sex by altering the chemical processes that occur in development. The cells bearing XX produce a secretion that differs in its physiological effects from that produced by the cells bearing XO. These two diverse secretions bring about the different developmental processes that result in the production of the two sexes.

As before seen, the difference in the secretions or hormones results in some way from the fact that the XX cells have two centres for the production of certain materials, while the XO cells have only one centre for the production of those materials. The different proportions of materials result in diverse chemical processes that give the two different hormones, and thus result finally in individuals of different sex.

MAY OTHER THINGS BESIDES THE CHROMOSOMES INFLUENCE OR DETERMINE SEX?

We have gone into the matter of sex determination primarily for the purpose of getting light on the method of action of the chromosomes. We have seen that the nature of the chromosomes present affects the chemical and physiological processes of development, and in this way determines which sex shall be produced.

This suggests certain other questions regarding sex that it will be well to examine briefly. It is commonly said that sex is determined by the X-chromosomes (a true statement so far as it goes); and from this it has often been concluded that sex cannot be determined in any other way. In particular it has been held that this excludes the possibility of the determination of sex by particular environmental conditions.

This conclusion is, however, quite without justification. The fact that chromosomes determine sex does not prevent other things from determining it. The chromosomes determine sex only in the sense that changing the chromosomes alters the sex. We have already seen that changing other things likewise alters sex. Changing the X-chromosomes alters sex; changing the autosomes alters sex; changing the germ glands present alters sex; changing the hormones present alters sex. It will be worth while to enumerate various different ways in which sex could be determined, in view of our knowledge of the method of action of the chromosomes. Some of the ways are known to be realized; others are not as yet known to occur.

SUMMARY OF POSSIBLE WAYS OF DETERMINING SEX

1. Selective action exercised on the germ cells containing particular chromosomes.In the organisms of Group I, there are two kinds of sperms; those containing an X-chromosome (which produce females), and those containing no X (which produce males). Any agent that would determine whether certain eggs are fertilized by an X sperm or by one without an X would determine the sex. There are a number of possibilities of this sort.

    a. It is known that the sperms containing an X-chromosome are in some animals larger than those without X. It may well be, therefore, that one kind of sperm is more vigorous and active than the other. This kind would then fertilize a greater proportion of eggs than the other, so that a majority of the offspring would be of one sex rather than the other.

    There are indications that something of this sort is true in certain organisms. Males and females are not produced in equal numbers in all organisms. The average number of males to every 100 females are for certain species as follows:

Man 105, Cattle 107, Pigeon 115, Gottus (fish) 188, Lophius (fish) 385, Squid 17, Octopus 33.

Such disproportion between males and females might also be produced in some of the other ways hereinafter mentioned.

b. It is known that in some organisms all the sperms of one type die. In a species of Phylloxera, according to Morgan, at a certain period in the life history all the sperms that do not contain X die. As a result, all of the offspring produced in the next generation are females.

c. If one of the two types of sperm is more resistant than the other, certain unfavourable conditions might injure or destroy one type and not the other. As a result, all or most of the eggs would be fertilized by the uninjured type of sperms. Then all (or most) would produce individuals of the same sex. If the sperms with X were the ones injured, then all the offspring would be males; in the alternative case, all the offspring would be females.

   Such effects might be produced by conditions of temperature, or chemical conditions, or by certain elements in the food of the parent. It would then be found experimentally that these conditions 'determine sex'. Whether such effects are actually produced is to be discovered only by observation and experiment.

d. The eggs might have such chemical or physical properties as to admit one of the kinds of sperm, not the other.

   This would determine what sex is produced. There are indications that something of this sort occurs in some organisms.

e. Certain chemical or physical conditions of the environment might alter the properties of the egg, making it admit one type of sperm, not the other. Such environmental conditions would then determine the sex to be produced.

2. Destruction or division of an X-chromosome. Anything that would destroy one of the X-chromosomes in a fertilized egg containing two X's would determine the sex: the individual produced would be a male instead of a female.

   There is strong evidence that such transformation of an individual that began as a female into a male (wholly or partly) occurs at times. There occur in insects individuals in which one half of the body is female, the other half male. In some animals it is easy to detect this, because the males and females differ in structure or programmes in all parts of the body. Such individuals are known as gynandromorphs. In Drosophila, individuals are found at times in which one half of the body, to the middle line, is of one sex, the other half of the other sex. Such individuals would be produced if, in an egg containing two X's, in the two-cell stage an X were lost from one of the two cells, but not from the other. The half retaining the two X's would remain female, while'the other half would become male.

   Loss of a chromosome in the process of cell division has been observed in certain cases. One of the chromosomes becomes entangled with the advancing cell wall, as the cell divides; it is thus removed from the cell. If this happened to the X-chromosome in one cell of the two-cell stage in Drosophila, a gynandromorph would be produced.

   Sometimes individuals are found in which three-fourths of the body is female, while one-fourth is male; or in other cases nearly the whole body is female, while a very little is male. Such conditions would arise if one X-chromosome were lost from a cell at a later stage of development.

   Gynandromorphs might also be produced if by some means the X-chromosomes of certain cells containing but one X could be caused to divide, while the other chromosomes remained undivided. There are no cases known in which this appears to be the method of action. Gynandromorphs occur in birds as well as in insects, but are not known in mammals.

3. Changes in the chemical processes occurring in the cell.We know that the chromosomes affect sex by influencing the chemical processes in the cells. Any agent that influenced these chemical processes, in the same or a reverse way, would determine which sex would be produced. As before seen, it would be necessary only to alter the balance of the chemical processes occurring. An agent that in Drosophila reinforced the action of the autosomes would turn the development in the direction of producing a male. An agent that increased the chemical action produced by the X-chromosomes would tend to cause the production of females.

   These considerations open the door to many possibilities; since in the chromosomes we are dealing with centres of chemical action, these might well be favoured or hindered by various conditions. Changes in the chemical processes might well be induced through the action of certain types of nutrition, through temperature changes, through reactions to various stimuli, and the like. Whether such conditions determine sex is to be discovered only by observation and experment; there is nothing in the known method of action of the chromosomes to prevent.

4. In some organisms the sex of the individual is known to be determined by the conditions under which it develops. Examples of this are the following:

   a. In a certain common mollusc called Crepidula plana, if the eggs are kept away from full-grown individuals, they develop into females. But if they are allowed to develop near to older specimens of Crepidula, they become males. In such eggs there is thus a very delicately balanced condition, which may be turned toward either sex by a slight change of conditions.

b. In a marine worm known as Bonellia, the female is a large worm with a proboscis. The male, on the other hand, is very small, and is parasitic within the body of the female (in the uterus). The eggs develop into small creatures that swim about in the water. If they find a female, they attach themselves to her proboscis and develop into males. But if they do not become attached to a female, they themselves develop into females.

   Here again there is a delicate balance as to sex: which direction development shall take depends on the external conditions.

DIFFERENT CONDITIONS AS TO SEX IN DIFFERENT ORGANISMS

There are many other conditions with respect to sex among different organisms. It will not be possible to deal with all these, but it will be worth while to enumerate some of the more important conditions.

1. Organisms with the two sexes in separate individuals, the sex being determined by differences in the chromosomes, as by the presence of one X or of two. It is mainly such organisms that have been dealt with in the foregoing pages. As we have seen, there are among these two groups, having diverse relations of chromosomes to sex. In Group I, the female has two X-chromosomes, the male but one; in Group II the reverse condition is found.

2. Organisms having both sexes united in one individual (hermaphrodites). The same individual produces both ova and sperms. This condition exists in many organisms, as in common snails. The relations of this condition to chromosomes is not clear.

3. Organisms in which the individuals are male during one period of their lives, female during another. Certain fish show this condition; it is indeed found in many animals.

4. Organisms that are diverse as to sex in successive generations. Among these are to be distinguished several different types, as follows:
   a. Some organisms are hermaphrodites in one generation, but have separate sexes, male and female, in a later generation. Certain thread worms (Nematodes) show this condition.
   b. Some have no sex in one generation, reproducing without sperm and ova, but in the next generation have the two sexes, or are hermaphrodites. This is the condition found in plants.
   c. Some are exclusively female in certain generations, reproducing by parthenogenesis, the eggs not requiring fertilization. In a later generation there are both males and females. Many Rotifera are of this type.

The relation to chromosomes of these diverse conditions as to sex remains for the most part to be discovered. But the great variety of conditions found among organisms must be kept in mind, in order that the role of the chromosomes may not be misunderstood. They determine the sex through the fact that they act chemically; they determine it by altering the chemical conditions. Other things may alter the chemical conditions too, and then these also may determine sex.

In general, the relations of chromosomes to sex, brought out in the present chapter, demonstrate that the chromosomes affect development and characteristics through the fact that they influence the chemical materials and processes. Other conditions may likewise influence these processes, so that the effects of chromosomes may be modified or nullified by other conditions.

Chapter 7

Relation of Genetic System to Characteristics

X-CHROMOSOME AS TYPE

In the preceding chapter it has been seen that differences in sex are produced by altering the chromosome combinations present in the fertilized egg, and this has brought to light some of the methods by which the chromosomes act on development and characteristics. We now Examine other effects of altering chromosomes.

We shall have to examine the effects of altering all the different classes of chromosomesýÿthe X-chromosome, the Y-chromosome, and the different autosomes. Each of these three groups (X, Y, and autosomes) gives rise, as we shall see, to a different type of inheritance. The simplest relations, from the experimental point of view, are presented by the X-chromosomes. We therefore deal first with these.

EFFECTS ON CHARACTERISTICS PRODUCED BY ALTERING THE X-CHROMOSOMES

The X-chromosomes take such a course in passing from generation to generation that it is possible to follow the descendants of a particular Xchromosome (that present in the original male parent, for example), knowing in which individuals they are present, in which they are absent. Furthermore, in certain individuals (males, in organisms of

Group I) there is but a single X-chromosome instead of a pair. These relations make it a relatively simple matter to discover the distinctive effects of a particular X-chromosome. The effects of X-chromosomes will therefore be dealt with somewhat fully as a type of chromosomal action. Later the action of the other chromosomes will be taken up.

METHOD OF INHERITANCE OF CHARACTERISTICS RESULTING FROM MODIFICATIONS OF X-CHROMOSOMES

The X-chromosomes, as we saw in earlier chapter pass always from the father to his daughters, not to his sons. The sons get their X-chromosomes exclusively from the mother. These peculiarities make it relatively easy to discover characteristics of individuals that depend on the kind of X-chromosomes they contain. We saw in our introductory account of the discovery of the materials of heredity certain effects of altering X-chromosomes. Sometimes, as there set forth, an X-chromosome of one of the parents is defective in such a way as to cause an abnormality or defect in all the individuals that receive this X-chromosome or its descendants. Such an abnormality was bar-eyej. Characteristics of this sort, that are manifested in every individual that has the defective X, are called dominant characteristics.

It was further set forth that there are other characteristics which result from defective X-chromosomes, but which are manifested exclusively in individuals that carry only X's that are defective; if a normal X is present also, the characteristic does not appear. Such a characteristic was white eyes. Seemingly the defective X fails to produce certain required materials. But, if a normal X is present, it produces the materials, so that no defect results. Such characteristics, that are manifested only if a normal X is not present, are called recessive.

As we shall see in detail, many different characteristics are known that are thus due to defects or alterations in certain X-chromosomes. Such characteristics show very remarkable rules of inheritance. A number of such characteristics were known in man before their relation to chromosomes were discovered; their method of inheritance was extremely puzzling. It will be worth while to observe the rules of inheritance of such characters, as they appear when the relation to X-chromosomes is not brought out. In man there has long been known a defect called haemophilia, which follows this method of inheritance. It is due to a defective X-chromosome, and shows itself in the fact that the blood does not coagulate on exposure to the air, so that if wounded the defective individuals are likely to bleed to death. The defective X-chromosome fails to produce certain materials that are required for coagulation of the blood. But if a normal X is present in addition to the defective one, the necessary material is produced and the blood coagulates normally. Haemophilia is thus a recessive character. Programmes-blindness is another recessive characteristic that is due to a defective X-chromosome.

The following are the rules of inheritance shown by such a recessive character resulting from a defective X-chromosome.

1. The individual affected by the abnormality is usually a male.
2. When such an affected male mates with a normal female, none of the children are affected. It appears as if the defect were not inherited.
3. But in the grandchildren the defect reappears; it has 'skipped a generation.
4. But not all the grandchildren have the defect. There is a peculiar distribution of the defect among them, as follows:
   a. None of the sons' children have the defect; nor does it reappear in any of their descendants. So far as the descendants through the sons are concerned, the defect has disappeared.
   b. But some of the daughters' children have the defect.

c. None of the daughters' daughters have the defect.

d. But some of the daughters' sons have it. On the average it turns out that about one-half of the daughters' sons have the defect.

5. In later generations it continues as a rule in the way we have described, that is:

a. Only males have the defect.

b. It skips a generation, and reappears only in part of the grandsons through the daughters.

c. It is thus inherited through normal females.

d. But it is never inherited through normal males.

The above is the usual course of inheritance; it holds for the great majority of cases.

6. But observers were perplexed by finding that in very rare cases all these rules are broken. The following were then observed:

a. The individual affected is a male, as before.

b. He marries a normal female; and now some of the children are affected.

c. But not all the children are affected; on the average about one-half of all.

d. Those affected include both males and females, in equal number. Thus here we find that the rule that only males are affected does not hold; females also may have the defect.

7. When now one of these rare defective females is mated with a normal male:

a. All the sons produced are affected.

b. None of the daughters are affected.

Thus 'the sons inherit from their mothers, the daughters from their fathers', in such a case.

One further important fact was observed:

8. If the mother and father are both affected, then all children are affected, including daughters as well as sons.

Imagine trying to get intelligible rules of inheritance from such a set of seemingly contradictory observations! Even the rules commonly followed seem arbitrary and incomprehensible. And at times they are all broken; sometimes one result is produced, sometimes another.

But when it was discovered that such defects are due to a defective X-chromosome, and that they appear wherever the defective X-chromosome has no normal one with it, all this was at once cleared up and became intelligible. This is revealed at once when we place in the diagram the X-chromosomes of the individuals, marking in a special way the X- chromosome of the individual originally affected and in the same way all the X-chromosomes derived from that chromosome. Every individual that contains only the defective X is itself defective, whether male or female. Every individual that contains in addition a normal X is without the defect. Such defects usually appear in males only, because males have only one X-chromosome; and if that is defective, the individual is defective. Females, on the other hand, have two X-chromosomes, and as normal X's are more common than defective ones, usually any female that has a defective one has a normal one also, and is therefore not personally defective.

But rarely it happens by chance that a female gets two defective X-chromosomes; then, having no other, she is personally defective. And in that case, since she always gives one of her X's to her sons and they have no other, her sons are all defective.

Indeed, when we recognize that such a characteristic as haemophilia or programmes-blindness is due to a defect in certain X-chromosomes, but is not manifested if a normal X-chromosome is present, we discover that all the results enumerated above, and illustrated in the diagram of are inevitable.

Many other recessive characteristics, as we shall see, are due to defects or modifications of X-chromosomes. Such characters are distributed to the offspring in the way illustrated above; they all follow the same rules of inheritance. They follow the defective X wherever it goes, being manifested whenever the defective X is the only kind of X that is present. Such characters, since they are commoner in males, are commonly called sex-linked characteristics.

In the case of certain serious defects, these rules are modified by the fact that individuals with two defective X's (females) cannot live and develop. Haemophilia, for example, is not certainly known to occur in females. Any female that gets two X's defective in this way simply does not live. Programmes-blindness, however, illustrates well all the rules set forth above, and the same is true for many such defects in animals.

ABNORMAL DISTRIBUTION OF X-CHROMOSOMES

What will happen if the X-chromosomes by accident become irregularly distributed? We know that sometimes the X-chromosomes are indeed irregularly distributed. Will the sex-linked characteristics continue to follow them; will the defective characteristics show the same irregular distribution as do the X's?

Such cases in great number have been fully observed. It is found that the sex-linked characteristics do indeed follow the X-chromosomes wherever they go. This proves conclusively (if there were any possible doubt in view of the extraordinary course normally followed by such characters) that it is indeed the X-chromosomes on which the characters depend. The matter is one of importance and interest, so that it will be worth while to examine carefully certain typical cases of the result of irregular distribution of X-chromosomes.

NORMAL DISTRIBUTIONS

Sometimes the body of Drosophila is yellow instead of the normal grey; this is a recessive character due to a defect in the X-chromosomes. Suppose that we mate together a female that has a yellow body and a male that has the normal grey programmes. The female has two modified X-chromosomes, which we may represent by XX. The male has its X unmodified; we may represent its cells as XY. Normally in forming the germ cells, the two X's of the female separate into different germ cells, and all the ova receive one of the modified X's. The normal male produces germ cells of two classes, X and Y. When the germ cell X from the female unites with Y from the male, sons are produced, with the constitution XY; while X from the female with X from the father gives daughters XX. Since the yellow programmes dependent on X is recessive, the daughters XX have the normal grey programmes, like the father. But the sons XY have only one X, and as this is the defective one, from the mother, the sons are yellow, like the mother. The normal result of such a mating therefore is that all the sons are recessive like the mother, all the daughters dominant, normal, like the father. This is what occurs when the X-chromosomes are distributed to the germ cells in the usual way.

ABNORMAL DISTRIBUTION

Sometimes, as seen on previous pages, in forming germ cells both X's of the female go together to a single ovum, while other ova receive no X. In a race discovered by L. V. Morgan the two X-chromosomes in the female were partly united, so that they almost invariably thus go together to one ovum, leaving other ova without an X. In this race the X-chromosomes were so modified as to produce the recessive yellow programmes described in the preceding paragraph. We therefore have an opportunity to determine the course of inheritance of the yellow and the grey programmes when the X-chromosomes are thus abnormally distributed.

Represent as before the X-chromosomes that produce the yellow programmes by X. Thus one set of ova receive XX, while another set receive no X. When the germ cells XX united with the germ cells Y from the normal father, there were produced daughters XXY, in which both X's came from the mother, and these daughters were yellow like the mother (since no normal X was present). Sons were produced by the union of ova that contained no X with sperms that carried X; this gave sons XO, with their single X from the normal father, instead of from the mother, as is usually the case. And such sons had the normal grey body programmes of the father, instead of the yellow body programmes of the mother, as happens normally.

Thus with a change in the distribution of the X's there is a corresponding change in the method of inheritance. If the sons receive their X from the recessive mother, as in the normal cases, they are recessive like the mother. But when, through non-disjunction, the sons receive a dominant X from the father, they are dominant like the father. Similarly, in the normal case the daughters receive a dominant X from the father and are therefore dominant like the father. But in cases of non-disjunction the daughters receive both their JPs from the recessive mother, and are therefore recessive like the mother. In sum, when the X's are normally distributed, such matings give 'criss-cross inheritance'; sons like the mother, daughters like the father. But when the X's are not so distributed, there is no criss-cross inheritance; sons are like the father, daughters like the mother. The dominant and recessive characters follow the respective X-chromosomes, whether these are distributed normally or abnormally.

Such experiments have been repeated many times, and with other sex-linked characters. Always the characteristics follow the distribution of the X-chromosomes, however these are distributed.

It may be concluded with certainty that it is the presence of X-chromosomes of a modified type that causes the appearance of the particular sex-linked characters that are manifested. The rules of distribution of sex-linked characters are the rules of distribution of the X-chromosomes.

DOMINANCE AND RECESSIVENESS IN SEX-LINKED CHARACTERS

As we have seen, and shall see further, many defective conditions are due to defects in certain X-chromosomes. These follow from generation to generation the distribution of these X-chromosomes. Most of these bodily defects are manifested only in individuals in which the defective X- chromosomes are the only kind present; that is, these defects are recessive.

But it is important to observe that in such cases the normal condition of the organism likewise follows the distribution of certain X-chromosomes. Haemophilia follows certain X-chromosomes. But in the same matings in which this occurs, the healthy condition of the blood follows certain other X-chromosomes. If a defective mother is mated with a normal father, the sons are defective because they receive only the mother's defective X-chromosomes. But in the same way the daughters are normal because they receive the father's normal X-chromosomes. In the normal individuals the X-chromosome plays a part, just as it does in the defective individuals. In the normal individuals it is healthy and supplies what is required for normal development, while in the defective individuals it fails to supply what is required.

In the case of most defects this normal condition is dominant over the recessive condition; that is, when both are present, the normal chromosome is the one that prevails in its effect on the individual. There are defective characters, however, in which the defective condition is dominant or partly dominant. In these cases, when a defective X and a normal one are present together, the defective one produces its effect, as in the case of bar-eye, described earlier. In many such cases the defect is less marked when a normal chromosome is present as well as a defective one; then the defective condition is said to be partly dominant.

Thus whenever an individual bearing defective X-chromosomes is mated with a normal individual, we have both dominant and recessive characters, the course of which may be followed in the later generations the dominant character being commonly the normal or usual condition. How this works out is shown in our next paragraphs.

TESTS FOR SEX-LINKED INHERITANCE

By observing the distribution of the dominant and recessive characters among the offspring of certain matings, a test for determining whether given characters depend on the X-chromosomes is supplied.

The test consists in making what are called reciprocal crosses.

1. On the one hand mate a dominant female with a recessive male;
2. Also mate a recessive female with a dominant male.

Represent a dominant X-chromosome by a capital X, a recessive X-chromosome by a lower-case letter x. In the group of organisms with which we have been dealing, the female has two X-chromosomes, the male but one. The daughters receive an X from each parent, the sons an X from the mother only.

A dominant female will be represented by XX, a recessive male by xo; similarly a recessive female is xx, a dominant male is XO. The two ma tings and their results will then be represented as follows:

1. XX by xo gives X + XO Dom. Mother Rec. Father Dom. Daughters Dom. Sons
2. xx by XO gives X + xo Rec. Mother Dom. Father Dom. Daughters Rec. Sons

Thus when the mother is dominant, the father recessive, all the children are dominant, like the mother. When the father is dominant, the mother recessive, the daughters are dominant like the father, the sons recessive like the mother; this is called 'criss-cross inheritance'.

Whenever reciprocal crosses give these results, it is certain that the two diverse characteristics (dominant and recessive) result from differences in the X-chromosomes of the two parents. 'Criss-cross inheritance' is particularly useful in showing at once that we are dealing with sex-linked inheritance. In every case where such results are produced, further tests show that the two characteristics follow in later generations the two different kinds of X-chromosomes wherever they go. It is mainly by the use of these tests that the many different characteristics dependent on diversities in X-chromosomes have been discovered.

SEX-LINKED INHERITANCE IN GROUP II.

Sex-linked inheritance as we have just described it was originally discovered in animals belonging to Group I, in which the females have two X-chromosomes, the males but one (with or without a Y- chromosome).

But when the tests we have just described were applied to birds and to certain other animals, it was discovered that there is a second group (Group II) in which it is the male that has two X-chromosomes, the female but one. In the common fowl, reciprocal crosses were made between 'barred' fowls (Plymouth Rock) and black fowls (Langshan). Here the 'barred condition was found to be dominant. The results were:

1. Barred father by black mother=Barred sons + barred daughters.
2. Black father by barred mother = Barred sons + black daughters.

Here the second mating gives 'criss-cross inheritance', showing that we are dealing with sex-linked characters dependent on differences between the X-chromosomes of the parents. But, in the first mating above, all the offspring are like the father instead of like the mother, which latter was the case in the matings on the previous pages. The father in this case has the dominant characteristic (barred), and all the offspring are dominant like the father instead of like the mother. When the mother is dominant ('barred') the result in the fowls is criss-cross inheritance; while, in the cases before described, criss-cross inheritance occurs when it is the father that is dominant. In fact, in the birds, the males and females simply exchange roles, as compared with their roles in the organisms of Group I. This appears clearly from the following comparison of results in the two cases:

Parents Offspring

Group I. Dom. Mother, Rec. Father

Group II. Dom. Father, Rec. Mother) A" offspring dommant Chromosomes ýÿ XX xo Xx ýÿ XO

1. Group I. Dom. Father, Rec. Mother:ýÿDom. Daughters, Rec. Sons Group II. Dom. Mother, Rec. Father:ýÿ Dom. Sons, Rec. Daughters Chromosomes ýÿ XO xx X* ýÿ xo

Since the roles of males and females are interchanged in the two groups, their chromosomal conditions must also be interchanged; that is, since in Group I the female has two X's, and in Group II the male plays the same role as the female of Group I, the male of Group II must have two X's, the female but one. The chromosomal conditions in the two groups are shown in the above tabulation, X signifying dominant, while x signifies recessive, and o signifies the lack of X (whether Y is present or not). Careful examination shows that this is the only way in which the results in Group II can be produced.

It was in the way illustrated above that the existence of Group II was discovered. Birds and certain moths show inheritance of the kind typical for Group II. In both these it has since been found under the microscope that the males have indeed one more chromosome than the females.

THE ROLE OF THE X-CHROMOSOMES IN DEVELOPMENT

A great many characteristics, of various kinds, have been found to be the result of modifications of particular X-chromosomes, and thus to follow in later generations the distribution of the descendants of those chromosomes. Examination of a number of these is desirable, both for their own importance and for the light they throw on the functions of the X-chromosomes.

In man, the following characteristics, among others, dependent on diverse types of X-chromosomes, are known from the fact that they show typical sex-linked inheritance.

Haemophilia.Lack of coagulability in the blood. This results from a serious defect in certain X-chromosomes. The existence of defective X-chromosomes having this result shows that the normal X-chromosomes play a part in supplying something necessary for producing normal blood that coagulates properly.

Programmes Blindness. The fact that defectiveness in X-chromosomes causes programmes blindness shows that the normal X's play a role in producing the normally functioning eyes.

NIGHT BLINDNESS.

Inability to see in a poor light.

Near-Sightedness of certain types.

Progressive atrophy of the muscles (Gower's disease).The fact that defects in X-chromosomes have this effect shows that the normal X's play a role in the normal functioning of the nerves and muscles.

A considerable number of other sex-linked characters are known in man. But man is a very unfavourable organism for the study of inheritance. Yet even the little that is known of sex-linked inheritance in man shows that the X-chromosome plays a role in many diverse bodily functions. For a more complete idea of its role, some organism that can be bred experimentally must be examined. For this purpose the fruit-fly, Drosophila melanogaster, is the best organism to select, since it has been studied more extensively than any other.

In the fruit-fly we find peculiarities of the following types that are dependent on alterations in particular X-chromosomes:

Many different eye colours. The normal eye programmes in this organism is a certain shade of red. Defects in different X- chromosomes result in producing, in the individuals that bear these defective chromosomes, many different shades of red, varying from deep red to a very light red, and thence to 'buff', 'ivory, and 'white'. More than a dozen different types of eye programmes are known to result from modifications of the X-chromosome. It is clear that the normal X-chromosome plays an important role in producing the normal eye programmes.

Structural peculiarities of the eye, such as 'bar-eye', 'facet eye', 'furrowed eye', and the like, are known to result from modifications of the normal X-chromosome.

Wing modifications. Many different conditions of the wings are known that depend upon diversities among X-chromosomes borne by different individuals. These affect all sorts of features of the wings: size, form, venation, function.

The normal X-chromosome obviously plays an important role in the full and normal development of the wings.

Body colours and markings.ýÿModifications of the X-chromosomes produce the body colours yellow, sable, tan, chrome, lemon, green, and the like, in place of the normal grey. Other changes in X alter the distribution of pigment on the body, giving the characters *dot\ and the like.

Body structure.A defect in X produces irregularities in the abdomen, known as 'abnormal abdomen'.

Legs.A defect in X-chromosomes causes abnormal development of the legs; some of them are wholly or partly reduplicated, so that the total number of legs is increased.

Bristles.Modifications in X-chromosomes result in various different changes in the bristles that are scattered over the body.

Many physiological conditions and functions likewise depend upon the X-chromosomes, since they are changed when these chromosomes are modified. Among these are the following:

Positive reaction to light. Drosophila with normal X-chromosomes fly toward a source of light. Those having defective X-chromosomes of the kind that produce a tan-coloured bodydo not fly toward a source of light.

Weakness and short life. Most of the, defective X-chromosomes that cause structural or other changes in the body (different eye colours, wing forms, and the like) produce likewise weakness and short life. The individuals bearing them are less resistant to bad conditions, and live for a shorter time than the individuals that bear the normal X-chromosomes.

Life and death. Some X-chromosomes have defects that are so severe that the individuals bearing them will not live and develop, unless there is present also a normal X-chromosome. Such defects are known as lethals. The presence of such lethal defects in the X-chromosome has the result that males bearing them die, since they have but one X; while females usually live, since they have an additional X that is often normal. If a mother that has one lethal X and one normal one is mated to a father that has a normal X, the result is that half of the sons which receive the mother's defective X fail to develop, while the other half of the sons,receiving the mother's normal X, live and develop. All the daughters live, since they all receive a normal X-chromosome from the father. The consequence is that in such families there are twice as many daughters as there are sons. In families of 100 to 200, as occur in Drosophila, this is very striking.

X-chromosomes are known to play in other organisms roles similar to those mentioned above for Drosophila, although in no other organism has the matter been so fully studied.

We may summarize what has been brought out above in the following statements:

1. Many defects that appear in individuals are the result of defects in the X-chromosomes that they bear.
2. Some of these defects are dominant; they are manifested in all individuals in which the defective X is present.
3. But most such defects are recessive; they are manifested only in individuals in which the defective X is the only kind present.
4. When both a normal X and a defective X are present, in most cases the normal X performs the required functions, so that the individual is not defective.
5. Since the female has two X-chromosomes, while the male has but one, such defects are more frequently manifested in the males. Usually one of the X's present in the female is without the defect, so that she is not defective. But when the male carries a defective X-chromosome, the defect is manifested.
6. It is thus advantageous, so far as the occurrence of defects is concerned, to have two X-chromosomes rather than one.

7. This gives the female a considerable advantage over the male in these respects. There are many different types of defects due to defective X-chromosomes; almost all of them are more common in males than in females. Some of these defects seriously injure the health, or even result in death, if the defective X is the only kind that is present. It is well known that, in general, the death rate is higher in males than in females; this is probably due to the fact that males have but one X-chromosome.
8. Among the different individuals of a species are scattered a great number of different types of X-chromosomes having different effects on development. Many of the diverse X-chromosomes are distinctly defective, causing personal defects in the individuals that bear them. Others are not defective.
9. Defects or modifications in X-chromosomes affect in different cases all parts and functions of the organism.

THE ROLE OF THE X-CHROMOSOMES IN DEVELOPMENT

What do the facts brought out in the preceding sections show as to the physiological functions of the X-chromosomes; as to their role in development?

For every kind of effect resulting from defects or modificacations in the X-chromosomes, there is a corresponding (opposed or diverse) action of normal or unmodified X- chromosomes. Since defective X-chromosomes of a certain type produce in Drosophila short, ill-developed wings, it follows that the normal X-chromosomes act on the development in such a way as to give long, well-developed wings.For if we substitute the normal X for the defective one, this causes the normal well-developed wings to be produced. Similar reasoning applies to all the defective conditions that result from defective X's; the normal X's act in such a way as to produce the corresponding normal conditions.

It follows therefore that the normal X-chromosomes play a role in the production of eye programmes and structure, body programmes and structure, in the development of the legs, the wings, the bristles, in producing normal health, resistance and vigour, and in various physiological processes. We know, further, that they play a most important part in determining sex, with all that this includes of structural and physiological influence.

Clearly therefore the X-chromosomes play a most important part in development. Their action is not limited to any single part of the body, nor to any single class of functions. They enter into the processes of development in such a way as to influence all parts of the body, all functions of the body. They begin to influence development very early, as was seen in considering the development of sex. And they play important roles in such relatively late processes as the production of eye colours. It can hardly be doubted that they enter into the developmental activities from practically the beginning, influencing all the later processes that occur.

AUTOSOMES AND T-CHROMOSOMES

In the two preceding chapters the X-chromosomes have been dealt with because of the great advantages they offer for experimental study. It was seen that they affect development in many ways, and that in a given species there are many different types of X-chromosomes in the different individuals, causing them to show different characteristics, structural and physiological.

Is the X-chromosome typical in these relations? Shall we find similar effects in the other chromosomes?

In addition to the X-chromosomes, there are the autosomes, which are present as a rule as several or many pairs. In man there are 23 pairs of autosomes: in Drosophila there are three pairs. There is, further, in many species the single Y-chromosome. The autosomes and the Y-chromosome are distributed from parent to offspring in characteristic ways, differing from the distribution of X- chromosomes. If they affect characteristics, these differences should appear in a different system of inheritance. We shall examine first the autosomes.

THE AUTOSOMES: TYPICAL MENDELIAN INHERITANCE

We have already seen that the autosomes play a role in development, in the fact that they help to determine the sex of the developing individual. In Drosophila, if the number of sets of autosomes, in relation to the number of X's, is changed, this changes the sex.

Do autosomes also affect other characteristics? Do the different pairs of autosomes have different functions? Are there diverse types of autosomes, with different effects, in different individuals of a species, as is the case in relation to the X-chromosomes? Are there dominant and recessive effects of autosomes, as there are of X?

If dominant and recessive characteristics depend on autosomes, then these characteristics must follow the distribution of the autosomes from parents to offspring, as the characteristics dependent on X follow its distribution. We must then examine the method of distribution of the autosomes, to see whether there are characteristics that show this method.

METHOD OF DISTRIBUTION OF THE AUTOSOMES

The distribution of the autosomes from parents to offspring is much simpler than that of the X-chromosomes. The main facts are as follows:

1. The autosomes are in pairs in all individuals. There are no individuals with unpaired autosomes, like the male of some organisms with relation to X.
2. Thus the two sexes are alike as to their autosomes.
3. In forming germ cells, one member of each pair of autosomes goes to each germ cell. The sperms and the ova are alike with respect to the autosomes.

Thus the fundamental rule with relation to the autosomes is that each germ cell carries one member of each pair of autosomes.

4. By union of two germ cells the pairs of autosomes are restored.
5. Thus each one of the offspring carries in each pair of its autosomes one autosome from each parent.

Now consider a single pair of autosomes, such as the pair marked AA in. Suppose that in different individuals the members of this pair of autosomes produce different effects, as we have seen to be the case with the X-chromosomes of different individuals. And suppose that, as in the case of X, one of these effects is dominant, the other recessive. How would such characteristics be inherited?

This question must be examined with care. For this purpose we may take as an example the different colours found in different varieties of peas, used in the classic experiments of Mendel. The yellow seed programmes of a certain variety is dominant, and results from the constitution of the autosomes of a certain pair. The green seed programmes of another variety is recessive, owing to a different constitution of the autosomes of that same pair. Call the two autosomes of that pair in the dominant yellow race by the capital letters AA, while the two autosomes that give rise to the recessive green programmes in the other race may be called aa. The dominant parent may be designated D, the recessive parent R,

When these two parents produce germ cells, each germ cell receives, of course, one chromosome for the pair. The germ cells from the dominant parent D have the autosome A, those from the recessive parent R have the autosome a.

The germ cells A, from parent D, unite each with a germ cell a from parent R giving offspring (zygotes) containing both A and a. These first generation offspring are commonly designated Fi (first filial generation).

And since the characteristic carried by the autosome A is dominant, all these Fi offspring manifest the dominant characteristic D (yellow, in case yellow-seeded peas are crossed with green-seeded peas).

Next observe the result of mating together two of these individuals Aa, of the Fi generation, giving the Fa (second filial generation).

Each parent Aa gives, according to the general rule, two kinds of germ cells, one kind carrying the dominant autosome A, the other the recessive autosome a. Each kind of germ cell from one parent unites with each kind from the other. That is, half the A germ cells from one parent unite with A germ cells from the other parent, half with a germ cells from the other parent; and the same is true for the a germ cells from the first parent. This gives in equal proportions Fa offspring (zygotes) as follows:

AA + Aa + a A + aa

But as the two combinations Aa and aA contain the same chromosomes, the proportions may be written AA + 2Aa-f-aa

The individuals of the constitution AA and Aa contain the dominant chromosome A; therefore they manifest the dominant characteristic, D. The individuals aa, having only the recessive chromosome a, manifest the recessive characteristic, R. Thus in this F2 generation, so far as manifested characters are concerned, the proportion of offspring are 3 Dominant to i Recessive

Thus if dominant and recessive characteristics depend on the two members of a pair of autosomes, when a dominant individual AA is mated with a recessive individual aa the general result is bound to be that the immediate offspring (Fi) all manifest the dominant characteristic, while in the Fa generation (produced by mating two individuals from Fi) there will be three times as many dominant individuals as there are recessives.

Now, this is the result that was discovered by Mendel in 1866; it formed the foundation for the working out of Mendelian heredity. The proportion 30 to iR in the Fa generation is commonly known as 'the Mendelian ratio'; it is one of the most characteristic proportions in this type of inheritance.

Mendel discovered that there are many different characteristics that yield this 3 to i ratio. If yellow and green peas are crossed, in the Fa generation there are 3 yellows to i green; if round peas and wrinkled peas are crossed, in F2 there are 3 round to i wrinkled; and so for many other characteristics in the pea plants with which Mendel worked.

And since the time of Mendel these same ratios have been found for thousands of different characteristics, in great numbers of different organisms, from the pea plant and Drosophila to man. The fact that inheritance gives these ratios is commonly spoken of as Mendel's law, and the type of inheritance which it exemplifies is spoken of as Mendelian inheritance. The type that gives in the Fa generation 3D to iR is the commonest method of inheritance, much commoner than the sex-linked method, in which the characteristics depend on differences in the X-chromosomes. This is obviously because autosomes are more numerous than X-chromosomes.

We find then that inheritance according to Mendel's law takes place exactly as it would if the characteristics concerned depended on the two members of a pair of autosomes. As will be shown later, it can be proved experimentally that such characteristics do indeed depend upon autosomes.

Thus the rules of Mendelism are the rules of distribution of the two members of a pair of autosomes, one having a recessive effect, the other a dominant effect. Mendelian heredity is heredity of characteristics that depend on differences in autosomes.

Mendel did not know of the relation of the characteristics to the chromosomes. He discovered, however, certain other important facts about this type of heredity, facts which have become clearly explicable since it has been learned that this type depends on differences in autosomes. These facts must be examined.

For this purpose it will be convenient first to define certain terms which are universally employed in dealing with inheritance.

By the union of two germ cells or gametes, as A and a, there is produced a zygote. If the two germ cells are alike with respect to the chromosomes or characteristics that they carry, the zygote so produced is spoken of as a homozygote; thus AA and aa are homozygotes. If the two germ cells that unite to form a zygote are unlike, as A and a, the zygote so produced is called a heterozygote, as Aa.

The two corresponding characters, one dominant, the other recessive, that are produced by the two differing members of a pair of chromosomes are spoken of as alternative characters, or alleles, or allelomorphs. Any given individual manifests one or the other of the two alternatives, not both. For the alternative characters we shall employ henceforth the term alleles.

The characteristic manifested by a zygote is its phaenotype^ while its actual constitution is its genotype. Thus, two zygotes AA and Aa both manifest the dominant character; the dominant character is the phaenotype, or the individuals are said to be phaenotypically dominant. But they are diverse in genotype; one is AA, the other Aa. In the case of a homozygote AA or aa, the phaenotype and genotype may be held to coincide.

Mendel crossed yellow peas with green peas, and found, as before stated, that in the Fi generation all the peas were yellow, so that yellow is the dominant allelomorph. In the F2 there were three times as many yellows as greens. The actual number of seeds in the F2 generation was in his experiments 8023, of which 6022 were yellow while 2001 were green, so that the ratio of dominants to recessives was, accurately expressed, 3.01 to i.

Mendel found that if he bred further the individuals of the F2 generation, some of these gave different results from others. In the pea plant it is possible to allow each plant to fertilize itself, so that both kinds of germ cells come from the same parent. When this was done with the individuals of the F2 generation, the following results were observed:

The green plants (recessive) gave only recessive green offspring (they 'bred true').

Of the yellow plants (dominant), one-third gave only yellow offspring (they 'bred true'), while two-thirds of them gave both yellow and green offspring, in the proportion of 3 yellows to i green.

The reason for this is clear when we consider the chromosomes on which the characters depend. If A represents the chromosome (autosome) that gives the dominant yellow programmes, while a represents the chromosome that gives the recessive green, the original parents were AA and aa, the Fi generation were all Aa, while the Fa generation showed the proportions AA -f 2 Aa + aa. The recessive individuals aa are homozygotes; their two chromosomes are alike, being aa. When germ cells are produced by these homozygotic individuals, all these germ cells must contain the chromosome a, and when two of these unite in self-fertilization, of course they produce in all cases again individuals aa. These are, of course, recessives (green), like their parents; that is, the green parents aa 'breed true'.

Similarly the dominant yellows, AA (constituting one-third of all the dominants), are homozygotes. Their germ cells all contain the chromosome A, and the offspring produced by union of such germ cells, in self-fertilization, are all AA, and so dominant yellow like the parents. One-third of all the dominants thus breed true.

But the individuals Aa are heterozygotic; that is, they have two kinds of chromosomes, A and a, in the pair. Thus they produce two kinds of germ cells, A and a, in equal numbers. And when these germ cells unite, they necessarily give, as we saw on an earlier page, offspring of three kinds, in the proportions AA + 2 Aa 4- aa Of these, AA and 2Aa will be dominants (yellow), while aa will be recessive (green).

Thus the homozygotes aa and AA breed true, while the heterozygotes, Aa, do not.

Furthermore, as Mendel discovered, if the heterozygotes Aa are mated with the recessive homozygotes aa, they give two kinds of offspring, Aa and aa; half (Aa) being dominant,

the other half (aa) being recessive. If the heterozygotes Aa are mated with the dominant homozygotes AA, the offspring AA and Aa are all dominant, though half (AA) are homozygotes, the other half (Aa) heterozygotes.

All these results are typical for the descendants of two parents different in a single pair of chromosomes, one pair being dominant, the other recessive. For reference it will be well to make a table of these results, characteristic for Mendelian inheritance:

THE RESULTS OF MENDELIAN INHERITANCE

Parents: AA and aa; both homozygotic. F i: Aa; all alike, dominant heterozygotes.

Fs: AA + 2 Aa + aa. Three-fourths of the individuals dominant, onefourth recessive. Half the individuals homozygotes (AA and aa), the other half heterozygotes (aAa).

Each type of homozygote bred by itself produces offspring like itself.

The heterozygotes Aa, when bred by themselves, produce again offspring

AA + 2 Aa + aa Aa mated with aa yields again Aa + aa Aa mated with AA yields again Aa + AA. All of these results are found to hold true for thousands of different characteristics.

We may therefore summarize the situation in the following general propositions:

1. When two different individuals are crossed, in many cases the differences between them follow the same method of distribution as do the two members of a pair of autosomes ýÿone being dominant, the other recessive.
2. This distribution is what is known as typical Mendelian heredity.

3. The two sexes do not differ in these respects. Each sex has the same proportion of dominants and recessives.
4. The rules of Mendelian inheritance are essentially the rules of distribution of the members of pairs of autosomes.

Characteristics that depend on two or more pairs of autosomes.We have thus far followed the distribution of characteristics that depend on differences in a single pair of autosomes. But most organisms have several or many pairs of autosomes: the fruit-fly has three pairs, man has twenty-three. Do some characteristics depend on differences in one pair of autosomes, others on differences in other pairs? Have the different pairs of autosomes thus different functions in development and inheritance?

We find that the different pairs of autosomes have indeed different functions, and that some characteristics depend on one pair, some on another. This is shown in many cases in which two parents differ in two diverse characteristics. One characteristic may then act as if dependent on one pair of autosomes, the other on another pair. As an example of this, we may take a case studied by Mendel in peas.

One of the original pea plants had seeds that were (i) yellow, and (2) round, or smooth. These, as it turns out, are both dominant characters. The other plants, crossed with those just mentioned, had recessive characters; the seeds were (i) green, and (2) wrinkled.

When these two kinds of pea plants were mated together, it turned out, as we shall see, that yellow and green behave in inheritances as if due to a difference in the two parents in the effects of a certain pair of autosomes; that is, they are alleles. Round and wrinkled behave as if due to a difference between the two parents in the effects of another pair of autosomes. We may therefore represent the two pairs of autosomes. Representing the dominant characters by capitals, the recessives by lower-case letters, we may in the diagram represent the chromosomes giving origin to the four characters as follows:

A=yellow a=green

B = round b=wrinkled

The parents will then be represented as shown in Fig. The germ cells from these parents contain, of course, one member of each pair, as shown in Fig. One parent produces gametes all of the constitution AB, the other gametes all of the constitution ab. When a gamete AB from one parent unites with a gamete ab from the other, all the offspring (Fi) thus produced have the constitution AaBa, as shown at Fi in Fig.. Since the dominant autosomes A and B are both present, all these individuals are dominant for both pairs of characters; that is, they all have seeds that are yellow and round, as Mendel found. But all are heterozygotes-for both of these characters.

Next we mate together two of these Fi individuals to give the F2 generation. In forming gametes the rule is, of course, that each gamete contains one member of each pair of chromosomes. But the two pairs, Aa and Bb, are distributed to the gametes quite independently. Thus, some get A and B, some A and b, some a and B, some a and b. There are thus from each parent four types of gametes, AB, Ab, aB, and ab, each type being represented by many gametes. The gametes of each of the four types from one parent unite in approximately equal proportions with gametes of each of the four types from the other parent. That is, gametes of the type AB unite in different cases with AB, Ab, aB or ab from the other parent; and so of each of the other gametes of the first parent. The result is to give a set of 16 combinations of gametes (some of which are alike, however, as will be seen). It is not necessary to carry further in our diagram the outlines of the chromosomes.

INHERITANCE DEPENDING ON TWO PAIRS OF AUTOSOMES

AB AB AB AB

AB(i) Ab(6) aB(7) ab(s)

Ab Ab Ab Ab

AB(6) Ab(2) aB(5) ab(8)

aB aB aB aB

AB($_7$) Ab($_5$) aB($_3$) ab($_9$)

ab ab ab ab

AB($_5$) Ab(8) aB($_9$) ab($_4$)

But, as will be observed, not all these 16 are diverse; thus ABAb is the same in constitution or genotype as AbAB. If we number the different combinations, and give the same number to those that are alike, as is done (in parentheses) in the table, we find that there are nine different combinations or genotypes. That is, two parents that are both ABab produce offspring of nine different constitutions or genotypes, in the relative proportions given in the table. When the parents differed in but one pair of characters there were but 3 different combinations. When there are two pairs the number is 3>or9-

These nine different combinations will manifest the dominant characters A or B, in all cases where an A or B is present whether present once, as A, or twice, as AA); they will manifest the recessive characters a or b in all cases in which a or b is present twice, no corresponding capital letter being present. Collecting together these different types, we find the different characteristics to be manifested in the following proportions of the zygotes:

9 AB + 3 Ab + saB +1 ab

That is, 9 out of 16 are dominant for both pairs of characters, 3 are dominant for A, recessive for b; 3 dominant for B, recessive for a, and i out of 16 is recessive for both. There are thus 4 different phaenotypes. In Mendel's experiment with peas, in which the two pairs of characters were (1) yellow and green; (2) round and wrinkled; in Fi the proportions were 9 that were yellow and round, 3 that were yellow and wrinkled, 3 that were green and round, and i that was green and wrinkled. These proportions hold generally when there are two pairs of characters connected with different pairs of autosomes.

It may be observed that the proportions of the individuals of different constitution or genotype in F2, descended from a cross between parents that differ in respect to two pairs of autosomal characters (so that the parents are AABB and aabb), can be found directly from the proportions for a single pair of characters. In Fa, for the single pairs the proportions are AA + 2Aa-haA and BB + 2Bb + bb respectively. If now we multiply together algebraically these two polynomials, we obtain the relative proportions for the different combinations produced when the two pairs are together. These proportions are the following:

AABB + 2 AaBB + aaBB + 2 AABb + 4AaBb +

(1) (7) (3) (6) (5)

2aaBb + AAbb + 2 Aabb + aabb

(9) (2) (8) (4),

To bring this out, the same numbers employed in table 2 are placed in parenthesis beneath each combination.

Certain other relations among the nine different types are important, since they hold for all cases of two pairs of characters that are dependent upon different pairs of autosomes. These are the following:

1. There are four combinations (Nos. 1, 2, 3, 4), constituting one-fourth of all the zygotes, that are homozygotes for both pairs of genes. The four homozygotes are diverse in their combinations and in the characters they manifest. They form in table 2 the diagonal from upper left to lower right. Any one of these when self-fertilized or mated with another like itself breeds true: that is, it produces offspring that have the same constitution as the parents.
2. Four out of the 16, or one-fourth of all the zygotes, are heterozygotes for both pairs of characters. They form the class numbered (5), and constitute in table 2 the diagonal from upper right to lower left. Any of these when self-fertilized or mated with another like

itself gives again the nine different types, in the proportions above set forth.

3. The one-half of all the zygotes, are homozygotes for one of the two pairs, heterozygotes for the other. Four are homozygotic for A and heterozygotic for B; the other four are heterozygotic for A and homozygotic for B. In these 8 there are but 4 diverse combinations or genotypes.

If any one of these is self-fertilized, it yields 3 diverse types with respect to the pair for which it is heterozygotic. Thus, the class (6), having the constitution AABb, yields when self-fertilized three combinations in the following proportions:

AABB + 2AABb + AAbb

The above relations hold for all cases in which two pairs of characteristics are dependent on differences in two different pairs of autosomes, whatever the method of designating the characters or chromosomes.

In a similar way, two parents may differ in three characters that are dependent on three different pairs of autosomes. Such parents could be represented as AABBCC and aabbcc.

Their progeny (Fi generation) would all have the constitution AaBbCc. Such Fi individuals of course produce a variety of germ cells, in accordance with the principles that each germ cell receives one member of each pair of autosomes, and that the different pairs of autosomes are distributed independently. Each parent AaBbCc thus yields the following 8 types of germ cells: ABC, ABc, AbC, aBC, Abe, aBc, abC, abc.

When each of the 8 kinds from one parent unites with each of the 8 kinds from the other, there are produced, of course, 64 groups, which can be arranged in a table like the 16 groups of table 2. If we classify the 64 individuals by the characters which they manifest, we find that there are 8 different phaenotypes manifesting respectively the following characteristics:

ABC, ABc, AbC, aBC, Abc, aBc, abC, and abc

That is, some are dominant for all three characters; some for a particular two and recessive for the other; some are dominant for a particular one and recessive for the other two; and some are recessive for all three. The different phaenotypes are present in different proportions; the proportions when a large number are examined are as follows:

2?ABC + gABc + gAbC + gaBC + $Abc + saBc + 3abC+iabc

The total number of different combinations (genotypes) that occur in such a case is 27, or 3. These combinations can be obtained in their correct proportions by multiplying together algebraically the expressions:

Similar but still more complex series are obtained if parents are interbred that differ in four or more characters depending on four or more different pairs of autosomes. If the number of pairs of autosomes is called «, the following relations hold:

If parents differ in n autosomes:

1. The number of diverse types of gametes produced by the individuals of the Fi generation is 2^n.
2. The number of different combinations (genotypes) in the zygotes of F2, formed by mating the 2^n gametes from Fi,iss^w.
3. The number of different phaenotypes produced in Fa is2^n.

These relations yield very great numbers of diverse combinations, in cases where the number of pairs of autosomes is large. Thus in man, with 23 pairs of autosomes, if all the autosomes differed in the two parents, the number of types of germ cells produced by the Fi individuals would be 2, or many millions, the number of possible diverse genotypes in F2 would be 3, or many billions, and the number of possible different phaenotypes would be 2. (As will appear later, the number of different possible combinations is greatly increased by the exchange of parts by the members of a pair of chromosomes.)

Thus typical Mendelian inheritance, dependent on differences between different pairs of autosomes, can be treated to a certain extent mathematically. The individuals in F2 and later generations occur in proportions represented by simple mathematical ratios.

It will be worth while to formulate the main facts that we have brought out as to autosomal inheritance:

1. Some single pairs of characters, dominant and recessive, follow in inheritance the method of distribution of a single pair of autosomes.
2. Some pairs of inherited characters follow the rules of distribution of two or more different pairs of autosomes, in the following particulars:
   a. One member of each pair goes into each germ cell.
   b. The members of the different pairs are distributed to the germ cells independently, so that all possible combinations containing one member of each pair are formed with equal frequency.
   c. Each type of gamete from one parent unites with each type from the other parent, the different matings being equally frequent.
3. These methods of distribution result, for the characters manifested in the Fs generation, in certain typical numerical proportions or ratios, known as Mendelian ratios.
4. This method of inheritance is known as Mendelism, or Mendelian inheritance. The rules of Mendelism are the result of the rules of distribution of pairs of autosomes.
5. Most characteristics of organisms are inherited in this way; many more than are inherited in the sex-linked manner. This is evidently because there is but a single pair of X-chromosomes, while there are several or many pairs of autosomes.

6. Thus it is clear that, among the different individuals of a species, there are many different types of autosomes (as there are different types of X-chromosomes), producing many kinds of characteristics.

THE T-CHROMOSOME

There remains another chromosome, not included in the autosomes: namely, the Y-chromosome. Has this functions similar to those of X and of the autosomes? Does it too play a role in heredity?

As we shall see, the Y-chromosome differs in many ways from the other chromosomes. Let us first summarize the main facts as to the occurrence and distribution of Y-chromosomes.

1. Some organisms have no Y-chromosomes. Among those are many insects the dog, the cat, the horse.
2. In some organisms it is small and appears to be degenerate. This is the case in many insects; also in man.
3. In some organisms the Y-chromosome is large, but of a different form from its mate X. Such is the case in Drosophila
4. In some organisms Y is not distinguishable from X inform or size, but experiment shows that it differs from X in function.
5. The Y-chromosome descends from father to son only, exclusively in the male line. Normally it never occurs in females.
6. Hence any characters that depend on the presence of a particular kind of Y will *(a)* be found in males only, and will *(b)* never be inherited *through* females (since Y does not pass through females).

THE T-CHROMOSOME IN DROSOPHILA.

The functions of the Y- chromosome have been more thoroughly studied in Drosophila than in any other organism.

Most of the facts as to this have already been brought out, in connection with our account of the functions of the X-chromosome. We may summarize as follows the known facts as to the functions of the Y-chromosome in Drosophila:

1. The Y-chromosome, although it normally occurs in males, is not required for the production of male individuals. Through non-disjunction, male individuals are produced that contain no Y.
2. But the presence of Y is necessary for the fertility of males.
3. Y may exist in females, as a result of accidental nondisjunction; it does not affect their sex.
4. During many years of intensive study no characteristics were found that follow the Y-chromosome.
5. But recently it has been shown that the presence of Y prevents the appearance of a certain recessive character ('bobbed') that is due to a modified X-chromosome. This modified X results in the production of shortened bristles on the body of the fly. But this peculiarity does not occur if a Y is present, so that Y has the dominant normal allele for 'bobbed'. Indications of certain other ill-defined functions of the Y-chromosome have of late come to view.
6. No diverse modifications or defects such as are known in great number for the X-chromosome are known in the Y-chromosome.
7. Thus in Drosophila the Y-chromosome appears to have little function; it seems nearly but not quite inert.

Active T-chromosomes in certain Organisms; Relation to Male Secondary Sex Characters.In certain organisms the Y-chromosome is more active and varied in its effects. In such cases it shows in different individuals diverse modifications, comparable to the diverse types of the X-chromosome in Drosophila, and these produce diverse inherited characteristics that appear in males only.

This matter is closely bound up with the question of what produces the diversity of secondary sex characters. In many animals there are special characteristics which, like the Y-chromosome, are found only in males. These are the male secondary sex characters: horns, mane, beard, greater size, diversities in form, and the like. Since these appear only in the male, the question arises as to whether they may be due to the presence of the Y-chromosome. With relation to this the following is the situation:

1. In most cases, the presence of male secondary sex characters does not depend on the presence of the Y-chromosomes. This is shown by the following:
   a. There are animals which have no Y-chromosome, yet have male secondary sex characters: for example, the dog, the cat, the horse; also various insects.
   b. In Drosophila, males that contain no Y (as a result of non-disjunction) have the typical male secondary sex characters.
   c. In some breeds of sheep the male has horns, while the female has none. It could be suggested that this is due to the fact that the male carries the Y-chromosome, while the female does not. But in other breeds, as Dorsets, both sexes have horns, although only the male has the T-chromosome. And in still other breeds, as Suffolks, neither sex has horns, although here as elsewhere the males have the Y-chromosome. If a female of the horned breed (Dorset) is crossed with a male of the hornless breed (Suffolk), in the Fi descendants the males have horns, although they receive their Y-chromosome from the hornless breed. They have inherited the horns through the mother. Thus it is clear that in these

cases the presence of horns is not due to the presence of the Y-chromosome. Extensive work on the inheritance of horns and horn- lessness in sheep, goats and cattle consistently gives results which show that the presence of horns is not due to the presence of the Y-chromosome, or of any special type of Y- chromosome.

In all these cases (a), (b) and (c) the presence of distinctive male secondary sex characters depends in some way on the fact that the male carries but one X-chromosome, and consequently produces the male hormone. Their absence from the female results from the fact that she carries two X-chro- mosomes, and consequently produces the female hormone.

2. But in certain fish, some of the male secondary sex characters apparently do depend on the presence of a Y of a particular type, for they occur only where that type of Y is present, and they can therefore not be inherited through the mother. That is, the Y-chromosomes of certain individuals or breeds are so modified as to produce certain definite characteristics in the individuals that contain them, while individuals that contain another type of Y do not show these characteristics.

An example of this is found in the small fish Lebistes. In a certain race, the males have a black spot at the base of the dorsal fin, while the females are without the spot. In another race of the same species neither the males nor the females have the black spot. Thus the females of the two races are alike in lacking the black spot, while the males differ. Is this difference the result of differences in the Y-chromosomes of the two kinds of males?

This may be tested by crossing members of the two races. In the spotted race, represent the X and Y-chromosome by capital letters; while in the non-spotted race, represent them by the small letters x and y.

A female of the spotted race, XX, is mated with a male of the non-spotted race, xy, thus:

Females Males
Parents XX xy
Non-spotted
Fi Xx Xy
Non-spotted

THE Y-CHROMOSOMES

This produces in Fi females Xx, and males Xy. Neither contains the Y from the spotted race, and neither is spotted. The spot is not inherited through the females of the spotted race.

Next mate the female of the non-spotted race, xx, with a male from the spotted race, XY, as follows:

Females Males
Parents xx XY
Spotted
Fi xX xY
Spotted

This yields in Fi females Xx, and males xY. These males carry the Y from the spotted race, and they are spotted. Thus far the spot goes with the Y from the spotted race.

This Y may be traced further, for example, by breeding together the non-spotted females xx with the Fi males xY, from the preceding cross, thus:

Females Males
Parents xx xY
Spotted
F2 xx xY
Spotted

This gives in the Fs females xx, and males xY. These males are again spotted; they carry the Y from the spotted race. Wherever this Y from the spotted race is found, the black spot appears; wherever the Y is from the non-spotted race it does not appear. It appears clear therefore that whether the black spot appears in the male depends upon the type of Y-chromosome that it carries. There are two types of Y-chromosome in the species, one producing the black spot, the other not producing it.

It has been shown by Winge that there are many other programmes markings in this fish that depend for their occurrence on the presence of a particular type of Y-chromosome. They occur only in the males, and only in males that have a certain type of Y. In certain other fish also the Y-chromosomes have been shown to be diverse in different individuals, giving rise to distinctive characteristics in the male.

In a certain plant, Melandrium, also, there are inherited characteristics that depend upon the presence of a particular type of Y-chromosome.

There are some indications also that in man the nature of the Y-chromosome affects certain characteristics. In a family described by Schofield, the male parent in the first observed generation had webbed toes. This was inherited in the male lines only by all the males for four successive generations. It thus followed exactly the distribution of the Y-chromosome from the original male parent. It seems probable therefore that it is due to a defect or modification in the Y-chromosome of that individual. Many other cases of webbed toes or fingers are known that occur in both sexes, so that they are clearly not due to the Y-chromosome.

On the whole, the situation as to the Y-chromosome and its relation to inheritance may be summarized as follows:

1. In most organisms Y is a degenerate chromosome, lacking most of the functions manifested by the other chromosomes.
2. In some organisms degeneration has gone so far that the Y-chromosome has completely disappeared.
3. In other cases it exists, but is very small.
4. But in some species the Y-chromosome has effects on hereditary characters that are similar to those

produced by other chromosomes. Different types of inherited characters result in such cases from diversely modified Y-chromosomes.

5. Characteristics dependent on the presence of a particular type of Y-chromosome are inherited in the male line exclusively.

THREE MAIN METHODS OF DISTRIBUTION OF INHERITED CHARACTERISTICS

There exist therefore three diverse systems of distributing hereditary characteristics, depending respectively on the X-chromosomes, the autosomes, and the Y-chromosomes. In each of these the rules of inheritance are given by the special method of distribution of the three types of chromosomes, taken in connection with the facts of dominance and reces- siveness. Inheritance which follows the distribution of the X-chromosomes is commonly called sex-linked inheritance. That following the distribution of autosomes is known as autosomal or typical Mendelian inheritance. Less important than those just named is Y-chromosome inheritance. All these methods are included under the term Mendelian inheritance, employed in a broad sense.

In some organisms still other methods of inheritance exist. In higher animals and plants most inheritance occurs in one of the three methods named in the preceding paragraph.

Chapter 8

Constitution of the Chromosomes

GENES AND THEIR RELATIONS TO CHARACTERISTICS

We have seen in preceding chapters that there are many different modifications of a particular type of chromosome, such as the X-chromosome, producing diverse characteristics, which follow in their inheritance the distribution of the chromosome that produces them.

Do all these diverse modifications affect the same part of the chromosome; do they each affect the entire chromosome? That is, does the chromosome act always as a unit, a unit that is diversely modified in different cases, so as to produce different results? Or is the chromosome made up of different parts, with diverse functions: one part being modified when the eye programmes is changed, another part when the form of the limbs or of the body is altered, and so on? Are diverse characteristics influenced by different parts of the same chromosome?

Facts have been discovered which show that the chromosome is not a simple unit, but is composed of different parts, with different functions. We shall examine some of the more important of these facts and their consequences, taking the X-chromosome as a type.

As will be recalled, a recessive character that results from a modified or defective chromosome is not manifested when a normal chromosome is present in addition to the modified one. If the cells of an individual of Drosophila contain an X-chromosome that is defective in such a way as to cause, when by itself, the wings to be rudimentary, but also contains another X-chromosome that is normal, the wings are not rudimentary, but of the full grown normal type. Or if one X-chromosome tends to produce the recessive eosin-coloured eye, while the other tends to produce the dominant normal red eye, the individual bearing the two has the normal red eye.

But what will happen if two X-chromosomes that are so modified as to give different recessive characters are brought together in the same individual? What happens, for example, if in the cells of the same individual there is one X-chromosome so modified as to produce rudimentary wings, and another that is so modified as to produce eosin eyes? This can be brought about by mating a male that has the X defective in one way to a female that has its X's defective in another way. For example, an eosin-eyed mother is crossed with a rudimentary-winged father. Their daughters then contain one X so modified as to produce eosin eyes, the other so modified as to produce rudimentary wings.

Such matings have been made thousands of times, with many different types of defective chromosomes. The general rule is that in such cases the offspring are quite normal; they have neither of the defects. The daughters have neither eosin eyes nqr rudimentary wings; they are fully normal as to both eyes and wings.

The same result may occur even when the two chromosomes produce defects in the same part or function of the body. Cross together in Drosophila two parents, one of which has rudimentary wings, the other having miniature wings, both parents thus strongly defective. The daughters produced, bearing both kinds of defective X-chromosomes, have fully developed normal wings. In the same way, if we cross eosin- eyed individuals with vermilion-eyed (both these colours being the result of defects in the X-chromosomes), the offspring that contain both kinds of defective X's have normal red eyes. Similarly, yellow body crossed with tan-coloured body gives individuals with normal grey body.

Why do two diversely defective X-chromosomes, when brought together in the same egg cell, yield offspring that possess neither of the defects, but are fully normal?

We know that when a normal chromosome is present in the same pair with the defective one, the individuals are normal. This suggests a possible explanation. Perhaps the defect is in only a part of the chromosome, and in a different part in the case of diverse defects.

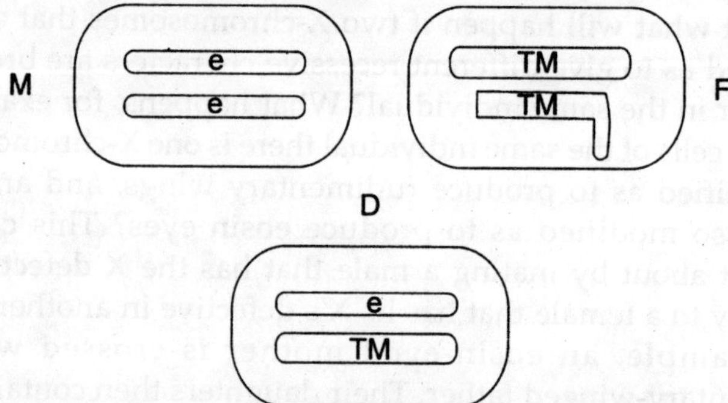

Fig .Diagram to illustrate the results of putting together two chromosomes that are defective in different ways. M, mother; F, father; D, daughter. The mother's X-chromosomes are defective in such a way as to produce eosin-coloured eyes instead of red eyes; the father's X is defective in such a way as to produce rudimentary wings. The daugh ters, receiving an Xfrom each parent, have both their X's defective, though in different ways. But they show neither of the defects, having normal red eyes and normal long wings.

Thus, if the defect that produces eosin eye is in the left half of the chromosome, while that which produces vermilion eye is in the right half, then when the two defective chromosomes are together, there is present a dominant normal left half and a dominant normal right half, so that the individual is normal.

According to this suggestion therefore:

1. The chromosome is composed of diverse parts and a certain part may become modified or defective, while the rest remains normal.

2. If one chromosome contains a dominant normal part corresponding to the recessive defective part in the other, the individual produced is normal, so far as this defect goes.
3. Hence when two chromosomes having recessive defects in different parts are brought together, the individual is a dominant normal.

This suggestion we must examine further to determine whether it agrees with the rest of the facts.

An important consequence is to be noted at once. A hundred or more diverse recessive characters are known that are due to modifications of X-chromosomes in Drosophila. The general rule is that when any two of these diversely modified X's are brought together in the same cell, the individual produced is normal (certain exceptions will be dealt with later). This would require, according to the suggestion that we are considering, that the X-chromosome is composed of a hundred or more diverse parts. Can this be accepted? There is further evidence on this point, set forth in the following.

Certain other discoveries were made as to the mating of parents having X's that are diversely modified. We know that when two recessive parents having the same modifications in the X's are mated, the offspring all show the same recessive character as do the parents. According to the suggestion we are considering, this is because the two X-chromosomes in the offspring have the same parts modified, so that no corresponding dominant normal part is present. Thus, if the two parents have eosin eyes, all the offspring have eosin eyes.

There have been discovered certain cases in which two X-chromosomes that are somewhat diversely defective do not yield normal individuals when brought together. This is true of certain recessive eye colours in Drosophila. If an eosin-eyed mother is crossed with a buff-eyed father, the daughters have both the modified X-chromosomes; yet their eyes are not of the normal red, but rather intermediate between eosin and buff.

According to the suggestion we are considering, this must be due to the fact that eosin and buff are due to recessive modifications of the same part of the X-chromosomes so that when the two are together no corresponding dominant normal part is present so that the normal red programmes cannot be produced. Several other eye colours have been found to act in this way in Drosophila. There is a series of colours, ranging from white through various shades up to near the normal red, each programmes being due to a modification of an X-chromosome. The different colours have been named white, ivory, buff, ecru, coral, eosin, tinged, cherry, blood. When any two of these are brought together in the same cell they give, not the normal red, but various intermediate colours. According to the suggestion we are considering, therefore, each of these must be due to a modification of the same part of the chromosome. Different X-chromosomes would have this particular part modified in different ways, so as to yield different eye colours. Is this a credible suggestion?

As we shall see later, it is possible to get other evidence as to whether all these modifications are indeed in the same part of the chromosome.

We have before us then a definite suggestion for examination. This suggestion is that the chromosome is composed of many diverse parts. Any of these parts may become so modified as to produce a recessive character. But if there is present in the cell that same part in the unmodified or dominant condition (in the other chromosome of the pair), then the recessive character is not manifested. How can we discover whether it is true that the chromosome is composed of diverse parts?

Evidence on this is obtained in another way. As we have seen, the sex-linked characteristics resulting from modification of an X-chromosome follow from generation to generation that X-chromosome. Observation shows an extremely important fact. Sometimes two or more diverse sex-linked characteristics follow the same X-chromosome. That is, a single X-chromosome is so modified as to bring about two diverse sex-linked characteristics in the individuals that carry it.

For example, certain X-chromosomes cause in the fruit-fly the eyes to be white, in place of the normal red. Certain other X-chromosomes cause the body to be yellow, in place of the normal grey. And sometimes we find individuals that have both peculiarities; they have white eyes and yellow bodies. Some of these individuals are males, having but one X-chromosome, so that the two peculiarities must both be due to the same X-chromosome.

And if we breed such male individuals, we find that both these peculiarities are indeed due to the same X-chromosome. Wherever the male's X-chromosome goes, these two peculiarities go with it. They follow that single chromosome for generation after generation, in the complicated course that is followed by the X-chromosome. And in other cases, more than two diverse characteristics are found thus to follow one single X-chromosome. Three, four, five or six have been found following one X-chromosome. And yet, in other cases, each of these same characteristics is connected with a separate X-chromosome and follows it in its distribution. That is, two or more characteristics may either be connected with separate X-chromosomes, or they may be all connected with the same X-chromosome. It is to be understood that these are experimental facts, demonstrated in thousands upon thousands of breeding experiments.

In view of these facts, an important question arises. When several characteristics follow a single chromosome, are they following different parts of it? Has the X-chromosome different parts that can be modified separately. Or is the chromosome an inseparable unit, so constituted, however, as to cause in the individual that bears it several different peculiarities?

Experimental breeding answers this question clearly. It is most important to observe what this answer is and how it is obtained, for it constitutes one of the foundation facts of heredity. The answer turns out to be, as we shall see, that the chromosome is indeed composed of different parts, which have different effects in development. These different parts, as we shall see, can be separated, each carrying with it its own peculiar effect. The diverse separable parts of the chromosome having diverse effects are what are called genes. Their existence is solidly demonstrated, and can be verified by anyone who will take the necessary trouble.

The type of experiment that proves this to be true is as follows: Select in the fruit-fly a male that has white eyes and yellow body. These characteristics are both connected with the single X-chromosome carried by the male, as can be proved by such breeding experiments as have been described above. Now mate such a male with a female whose X-chro- mosomes are of the usual type, so that she has the dominant normal red eyes and grey body. Their daughters (C) get, according to the usual rule, one normal X-chromosome ('red-grey') from the mother, one modified 'white- yellow chromosome from the father. In the cells of these daughters (C), therefore, these two diverse kinds of X-chro- mosomes are together side by side in the same cell. And while they are together, as the results show, the two may exchange parts, so as to form a new combination of characteristics, thus proving that each is made up of diverse parts.

This is discovered in the following way. Such daughters carrying the two types of X-chromosomes, are mated with normal males (D), having normal X-chromosomes, and therefore having red eyes and grey body. Observe the sons that they produce (E, F, G, H). These sons, as we know, receive their single X-chromosome from their mother only, getting only the Y from the father. The two X-chromosomes of the mother separate, as we know, into different eggs. The eggs that contain the normal or 'red-grey' X-chromosome, when fertilized by a Y sperm of course produce sons with red eyes and grey body.The eggs with the recessive 'white-yellow' X-chromosome (which came from the original male produce sons with white eyes and yellow body (F). Most of the sons produced are of these two types, in equal numbers. But, in addition, there are a few sons (about one per cent of all) that have a new combination of characteristics. There are a few sons that have red eyes and yellow body. And there is an equal number that have white eyes and grey body (H). It is clear that in the single X-chromosome of these sons a new combination has been made, composed of a part of the X-chromosome of the original father (B), and a part of the X-chromosome from the original mother (A). The mother's X and the father's X have exchanged parts, in some cases, while together in the cells of the daughters; the exchange has occurred in about i to i£ per cent of them, not in the others. The newly combined chromosomes behave in later generations just as did the original combinations. The new combinations (red-yellow and white-grey) follow in later generations particular single X-chromosomes.

In this formation of new combinations, it is clear;

1. That two characteristics that were connected with different X- chromosomes (as red eyes and yellow body) later become combined with a single X-chromosome;
2. That characteristics that were first connected with a single X-chromosome (as white eye and yellow body) have later become separated, so as to be connected with two different X- chromosomes. It is clear, therefore, that the white and the yellow are connected with different parts of the chromosome, and that these different parts can be separated.

This same process of forming new combinations of parts of previously existing X-chromosomes has been found to occur with any of the other hundred or more diverse kinds of X-chromosomes that are known to occur in Drosophila. The process of exchanging parts of the chromosomes is commonly spoken of as 'crossing-over' of a part from one chromosome to another. So remarkable a process has of course attracted the attention of many investigators, and it has been studied in hundreds of thousands of cases. Details concerning it we take up later.

All this answers the question, proposed earlier, as to whether the chromosome is a simple unit, or is composed of a number of different parts with diverse functions. The chromosome does indeed consist of diverse parts that may become separated. And it is clear that these diverse parts have different functions, different effects on development. One part of the X-chromosome of the original father, in the experiment we have described, causes the eye to be white; another part causes the body to be yellow; and these parts can be separated. In the original mother, one part of the chromosome caused the eyes to be red, another part caused the body to be yellow; and these parts also can be separated. Different parts of the chromosome affect different characteristics.

The chromosome may by means of radiation be broken into two or more visible pieces, and that the different pieces produce the diverse characters that are ordinarily connected with the single chromosome. The relations shown in the exchange of parts by two chromosomes, described above, are fully confirmed by the results of physically breaking the chromosome into pieces.

The separable parts of the chromosome, with diverse functions, are known as genes. It is important to observe that the existence of the genes, as separable parts of the visible chromosomes, is definitely proved by experimentation. Genes are not mere hypothetical units with imaginary properties, as is sometimes asserted. Genes are separable parts of the chromosomes, having diverse effects on development; their existence is solidly demonstrated, and can be verified by anyone who will take the necessary trouble. There is of course room for theorizing, in the present state of our knowledge, as to the precise nature and action of genes, but this does not touch the fact of their existence and effects.

How many separable parts are there in the chromosome? That is, how many genes are there? And how are they arranged; how are the parts put together to form the chromosome? And how is their number and arrangement discovered?

These questions are answered through an examination of the processes of exchange of parts of the chromosomes, in the case of different sets of characters. The evidence is somewhat complex, although entirely clear in its bearings on the questions once it is completely mastered. It will be possible to present only some of the main features of the evidence. If this is followed with care it will give an understanding of some of the essential features of the materials of heredity that can be obtained in no other way. The relations to be brought out are fundamental for a correct understanding of heredity; they should on no account be neglected by the student.

Go back first to the experiment.In this experiment, a 'red-grey' X-chromosome from the mother (A) and a 'white-yellow X-chromosome from the father (B) were brought together in the cells of the daughters (C). There, as is shown by the nature of the sons (E, F, G, H) produced by these daughters, in a small proportion of the cells the two chromosomes exchange parts, so that sons with four different types of X-chromosomes are produced, in place of the two original types that were present in the cells of their mothers (C). The proportion of the cells in which exchange has occurred is directly shown by the proportion of the sons that have the new combinations, red-yellow and white-grey. When this experiment is performed with thousands of individuals (as has often been done), the percentage of the different combinations among the sons is found to be approximately the following:

Red-grey 49-4
White-yellow 49-4
Red-yellow 0-6
White-grey 0-6

The new combinations are the Red-Yellow and the White-Grey, and together they form about 1-2 to 1*5 per cent of all. That is, exchange or crossing-over has occurred in about i»5 per cent of the cells of the mothers (C), while in the other 98-5 per cent it has not occurred.

This proportion in which exchange has occurred is known as the exchange ratio, or as the cross-over ratio. This exchange ratio plays a most important role in the evidence on the questions with which we have to deal.

CONSTANCY OF THE EXCHANGE RATIO FOR A GIVEN PAIR OF CHARACTERISTICS

Approximately this same exchange ratio is produced whenever this particular mating is made in a large number of cases. If by such matings a thousand sons are produced, and this is repeated, in each 1000 there will be about 985 to 988 showing the original combinations, about 12 to 15 showing the new combinations. If the original parents show the reverse combinations (red-yellow and white-grey), then the new combinations red-grey and white-yellow will appear in about 12 to 15 cases out of each 1000. Thus for a given pair of characters (as white-yellow) the exchange rate is approximately constant.

DIVERSE EXCHANGE RATIOS IN DIFFERENT PAIRS OF CHARACTERISTICS

If, in place of the characteristics we have been considering, some other pair of recessive characteristics is crossed with normals, a different constant exchange ratio is produced. Thus, if individuals with yellow body and bifid wings (characters connected with the X-chromosome) are mated with normal individuals (having grey body and normal wings), and the second generation is obtained in the way illustrated in the sons show an exchange ratio of about 5-3 per cent. And this ratio is produced, approximately, whenever this pair of characteristics is tested.

A SERIES OF DIVERSE EXCHANGE RATIOS

Still other pairs of characteristics give still different exchange ratios. If matings are made using many different pairs, many different exchange ratios are produced. It is particularly instructive to examine the ratios for a series of pairs, one member of which is the same in each case.

The exchange ratio of the character white eye with many other characteristics has been determined, but the above list gives typical results. There is a whole series of diverse exchange ratios, varying from about one per cent to nearly 50 per centýÿeach ratio being approximately constant when the given pair of characteristics is employed.

Why does there exist such series of diverse ratios as is illustrated above? When there is an exchange, the two parts of the given chromosome that produced the two characteristics become separated from each other; they break apart. If the exchange ratio is low, they break apart only in rare cases; if the exchange ratio is high, they break apart frequently. The question then becomes: Why do certain parts of the chromosome break apart frequently, others less frequently, in the way shown by the series of exchange ratios?

In judging of this, there are certain important facts to be considered:

1. It can be proved experimentally that the exchange of parts occurs at the time of the last two divisions in forming germ cells. (This is proved by subjecting the individuals at different periods in their lives to high temperatures, which change the frequency of exchanges. This result occurs only if the subjection to high temperatures occurs at the time mentioned.)
2. At this period the two chromosomes are side by side, in conjugation.
3. And each is seen to be a slender elongated thread, having thickenings at intervals, in linear series.
4. The breaking apart and exchange of parts therefore occur in these long slender threads.
5. Possibly, therefore, the differences in the frequency of separation of different parts of the thread are due to the fact that some parts are close together on the thread, others far apart. If the chromosome breaks in but one or two places, as appears to be the case, parts that are far apart would become separated more frequently than parts that are close together.

Represent the chromosome by a line on which there are consecutive thickenings, A, B, C,... to H, representing genes. Then if a break occurs anywhere in the chromosome it will separate the extreme genes A and H. But to separate any two adjacent genes, as E and F, a break must occur precisely in the short stretch that lies between them. Any break that separates two genes that are near together will separate others lying farther apart. The result of these relations is that the farther apart in the series two genes lie, the greater would be their frequency of separation, and consequently their exchange ratio.

The suggested answer to the questions is then as follows: The exchange ratios are different for the different pairs of characters, because the genes to which these characters are due are at different distances apart on the thread-like chromosome. Two genes that are close together give a small exchange ratio; two genes that are farther apart give a larger exchange ratio. The greater the ratio the farther apart are the genes. These relations therefore yield a series of diverse ratios.

Is this suggestion borne out by the known facts as to the chromosomes and as to the crossing-over of different genes? Let us examine the facts from this point of view.

If we think of white as at one end of the series, then the genes would be in the order: white, abnormal, bifid, club, vermilion, miniature, rudimentary, bar.Certain consequences of this idea appear at once, and these may be used for testing whether the idea is adequate.

1. Since the size of the ratio depends on the distance apart of the genes, two genes that have nearly the same ratio with another must be at about the same distance from it. Thus vermilion, with a ratio of 30-5 from white, must be at nearly the same distance from white as is miniature, since the latter has a ratio of 33-2. If white is at the end of the series, therefore, vermilion and miniature must be close together. And if they are close together, they must have with each other a small exchange ratio. This can be tested; it has been done on a large scale. The exchange ratio

for vermilion-miniature is found to be indeed small. This same relation is found for other characters that have similar relations with white. A large number have been tested, in addition to those mentioned in the list. In all cases, characters that have nearly the same ratios with white have together small ratios; while characters that have very different ratios with white have together large ratios. This agrees completely with the idea that we are testing. All the relations found between the ratios are such as are to be expected if white is near one end of the chromosome, and the other genes in the list are so placed in the linear series that the near ones give small ratios, the more distant ones large ratios.

Of course, if the gene for white were not near one end of the chromosome, two other genes might be at equal distances from it, but on opposite sides of it, so that they would be far apart, and would therefore together have a large exchange ratio. All the evidence, however, is that the gene for white is near one end of the chromosome (though not quite at its tip), There is evidence, however, that the gene yellow, which has with white the ratio 1*5, lies on the other side of white, close to the very tip of the chromosome.

2. But if the ratios of the other genes are taken with vermilion, for example we find exemplified the relation just mentioned: genes that have nearly the same ratio with it may behave as if they were on opposite sides of it, and so have together a large exchange ratio. In all these respects the relations are what is to be expected if the genes are in a linear series, and their distance apart determines the magnitude of the exchange ratios between them.
3. A most important consequence of the idea that the genes are in a linear series, and that crossing-over is due to breaks and exchange of parts in the thread-like chromosome, is as follows. When the chromosome breaks in such a way as to separate, into

different chromosomes and different individuals, two genes such as club and miniature.

Fig. Diagram to illustrate crossingover. In A, the paternal (p) and maternal (m) chains of genes. In B (a later stage], a part of the chain p (white) has become united with a part of the chain m (black); the two have exchanged part of their genes.

Then all the genes on one side of the break must go together to one individual, while all those on the other side of the break must go together to the other individual. Thus, in case of a break between club and vermilion the genes for yellow, white, abnormal and bifid would go with club to one individual, while the genes for bar, rudimentary and miniature would go with vermilion to the other individual. Crossingover would therefore not be confined to single genes, but would occur in blocks; and the genes that go together could be predicted from a knowledge of the exchange ratios of the genes.

This furnishes a crucial test of the theory, for if the genes are not in a linear series this relation is not to be expected. Examination shows that the requirement set forth in the preceding paragraph is indeed fulfilled. Crossing-over actually does occur in blocks of genes, those shown to be near together in the linear diagram going together to one individual; while a block of those that are together in a distant part of the series go to the other individual.

As to the intimate processes that occur in the chromosomes, when exchange occurs, the following facts are known:

a. The exchange of parts occurs while the chromosomes are conjugating in the process of forming germ cells.

b. They are then in the form of long threads, with corresponding thickenings on the two threads. These thickenings probably show the location of genes.

c. These threads become closely united; they are seen under the microscope to be intimately intertwined.

d. Later they separate, and study of the characteristics of the individuals produced from the germ cells, as described earlier, shows that at a certain point the original threads have broken, so that part of the thread p has become united with the remainder of thread m, as shown at B in Fig. 34. Thus a consecutive series of genes from chromosome p has become united with a similar series from chromosome m, as shown in the Fig.

Thus far, all the relations we have examined agree with the idea that the genes are arranged in the chromosome in a consecutive series, and that crossing-over occurs by break and exchange of parts in two such threads.

Certain relations brought out on earlier pages furnish further tests of this theory.

4. It was seen that when two chromosomes that are defective in different ways are united in the same cell, by mating parents bearing the two kinds of defects, the offspring produced are usually dominant normal. The reason suggested was that the two recessive defects are in different parts of the two chromosomes. Therefore each chromosome supplies a dominant normal part or gene corresponding to the recessive gene of the other, so that the individual is normal. In view of the known fact that there are

many genes in the chromosomes. If a and k are the two defective genes, chromosome 2 supplies a dominant gene A corresponding to the defective gene a, while chromosome i supplies a dominant gene K corresponding to the defective gene k.

Note now the relation of these facts to the exchange ratios.

CONSTITUTION OF THE CHROMOSOMES

If two chromosomes with defective genes yield normals, the defective genes are conceived to be in different parts of the chromosomes. They would therefore be at different distances from any third gene. And since the exchange ratios depend on distances, they should give different exchange ratios with that third gene. The gene k would give with T a smaller ratio than would the gene a. In general, when two recessive genes in different chromosomes yield dominants, the two recessive genes should have diverse exchange ratios with any third gene.

Is this in fact the case? Experimentation shows that it is. For example, in Drosophila, vermilion eye and eosin eye mated together give normals. The yellow- vermilion ratio is 34.5; the yellow-eosin ratio is 1-5. It is clear, therefore, that vermilion and eosin are in diverse parts of the chromosome.

This turns out to be a general rule. When two recessive genes in different chromosomes give when brought together a dominant or normal individual, these two recessive genes have diverse exchange ratios with any third gene, showing that they are in different parts of the chromosomes.

This relation becomes of particular significance when it is contrasted with another, to be taken up next.

5. In some cases, two diverse defects in different chromosomes do not yield normals when the two chromosomes are brought together in the same individual. Instead, they yield a defective condition that is usually intermediate between the defects due

to the chromosomes taken separately; or they yield a condition that approaches one of these defects more closely than the other.Thus when the chromosome that produces eosin eye is brought together with that which produces coral eye, the two give, not the normal red eye, but an eosin-coral eye.

This, it was suggested, is because the two recessive defects must be at the same point in the chromosome; they must affect corresponding genes of the two chromosomes, two members of the same pair. In consequence, when the two are together both genes of that pair are defective; no corresponding normal gene is present, so that a normal individual is not produced.

This situation again furnishes a test of the idea that the genes are in linear order. For if the two defective genes are indeed in the same gene pair then the two must, according to the linear theory, be at the same distance from any third gene, as A or T, and therefore they must give the same exchange ratio with any third gene.

This can, of course, be tested experimentally in such an organism as Drosophila; it has been done in many cases. In the example we have mentioned, eosin eye and coral eye both give with yellow body the exchange ratio 1-5 per cent. It is therefore true that they are located in the same part of the chromosome. The two also give the same ratios when tested with any other gene: for example, their exchange ratio with vermilion eye is about 33 per cent. There is a whole series of eye colours in Drosophila that acts in the way just described. Any two placed together give, not the normal eye programmes, but an intermediate condition. And all of them have the same exchange ratio with any third gene. Their exchange ratio with yellow body is 1-5 per cent, their common exchange ratio with vermilion eye is 33 per cent. They are all therefore different modifications or defects of the same part of the chromosome, the same gene. This series of eye colours has been mentioned before; it includes white, ivory, buff, ecru, coral, eosin, tinged, cherry, blood.

Such a series of diverse modifications of the same gene is commonly spoken of as a set of multiple allelomorphs or alleles. Many such sets of recessive multiple alleles are known in different organisms. They all show the same peculiarities. Any two (in different chromosomes) can be brought together into the same individual, giving, not the normal dominant characteristic, but a mixed or intermediate condition. And all the members of any such set have the same exchange ratio with any third gene, indicating that all are due to modifications of the same gene.

All of this agrees, as will be seen, with the theory that the genes are arranged in a consecutive series, and that different exchange ratios are the consequence of the different distances between genes that are tested. Hence, characteristics that are due to modifications of the same gene have the same exchange ratios with any other; while characteristics that are due to modifications of different genes have diverse ratios with any other.

6. There is a further important fact which again tests the adequacy of this idea of the arrangement of the genes. Two genes that are in different parts of the two chromosomes of the same pair can be brought by crossing-over into the same single chromosome.Thus, by a break and exchange of parts occurring in the region between the gene pairs aA and Kk, the two genes a and k would be brought into the same chromosome. On the other hand, two genes that are in the corresponding parts of the two chromosomes of a pair, as for example e and e′, cannot be brought by crossing-over into the same

chromosome, because the break is always at the same point in the two chromosomes of a pair.

The facts shown in breeding experiments agree precisely with these relations. Any two recessive genes that yield dominants when brought together into the same cell must, as we have seen, be at different regions of the two chromosomes and it is true that any two such recessive genes can by exchange be brought into the same chromosome. Thus, in the X-chromosome of Drosophila, if white eye is originally in one individual, yellow body in another, each connected with an X-chromosome, the two may be brought by crossing-over into the same single X- chromosome. Thus they can both be present in a male which has but one X-chromosome in his cells. Similarly, eosin eye and vermilion eye can be brought together into the same X-chromosome. In general, it is true that any two recessive genes that yield dominants, when brought together into the same cell, can, by crossing-over, be brought into the same chromosome (where of course they occupy different places).

On the other hand, it is an experimental fact that recessive genes that do not give dominants when brought together into the same zygote (and therefore must be at the same location in the two chromosomes) cannot be brought into the same chromosome. They are thus never present in the same male, since the male has but one X. Thus, eosin eye and coral eye cannot both be present in the same X-chromosome, though eosin eye and vermilion eye can. These, it will be understood, are experimental facts; many thousand experiments have given these results. Of the entire series of multiple allelomorphs, etc., no two, it is found, can ever be brought together into the same X-chromosome; no two can ever be present in the same male, though any two can be present in the same female.

This again of course agrees with the conception, originally based on other grounds, that all these are modifications of the same gene. When the chromosomes conjugate, the corresponding parts or genes of the two lie opposite one another, and the break which results in exchange of parts occurs at the same point in the two. Thus two genes of the same pair are not brought together into the same chromosome.

7. The validity of the conception that the genes are arranged in a chain can be tested by breaking the chromosome, through the use of radiation. Here we give only a general statement of the relation of the broken chromosomes to the linear arrangement of the genes. By the methods already described, the order of the genes in the chromosomes is determined, so that maps may be made showing this order. In Drosophila it is shown in this way, for example, that the genes producing yellow body, white eye, abnormal abdomen and bifid wings are near one end of the X-chromosome, that those for vermilion eye, miniature wing and furrowed eye are near the middle of the chromosome, and that those for bar-eye, cleft wings and bobbed bristles are near the opposite end of the X-chromosome from that which carries the genes for yellow body and white eyes. When individuals are subjected at the proper stages of development to X-rays, the chromosomes in their germ cells are frequently broken, and the broken chromosomes can be detected under the microscope in the cells of the offspring of the radiated individuals. Sometimes thus a small piece of one end of an X- chromosome is broken off; it may then become visibly attached to one of the autosomes. In such a case the characteristics dependent on the genes in this piece of X change their method of inheritance in later generations; they are now distributed to offspring by the method of autosomal inheritance,

instead of by sex-linked inheritance; for now they are attached to an autosome. It is therefore possible by study of the inheritance in later generations to determine what genes are present in the piece of X that is broken off. Such determinations have been made in many cases.

The results of these studies show that the order of the genes is correct. When a small piece of the end of the X-chromosome is broken off, such study as we have described shows that it contains the genes for yellow body, white eyes, and the other genes near the tip of the chromosome. If a longer piece of X is broken off, it contains more of the genesýÿthose extending farther toward the middle of the chromosome. Such studies, with the X-chromosome and with the autosomes, have been made on a large scale in Drosophila; they confirm throughout the order of the genes shown on the maps, the order originally determined by study of the different exchange ratios of the genes, as described in the foregoing pages.

Thus, in sum, all the tests to which the matter has been subjected agree with the conception that the chromosome is composed of separable genes, having different effects on development and characteristics; that these genes are arranged consecutively, like beads on a string; and that the different exchange^ratios of different pairs of genes are due to the different distances of the genes one from another.

This conception is subject to test in many other ways that cannot be described here. The consecutive arrangement of the genes results in many diverse mathematical relations, which must hold in the results of experimentation, if the consecutive order is correct. It may be stated that the results of literally hundreds of thousands of experiments show that these relations do hold, and they agree with and confirm this consecutive arrangement of the genes. Furthermore, the chromosome is seen under the microscope to be, at the time when crossing-over occurs, a linear structure with consecutive thickenings in other words, just such a structure as the results of experiments show it to be.

It may, therefore, be considered established that this conception is correct. The arrangement of the genes may therefore be represented. The genes in any zygote form two series, one series derived from the father, one derived from the mother. The two series contain corresponding genes. Thus the genes are in pairs, each pair containing one gene of paternal origin, one of maternal origin. The two genes of a pair have corresponding functions, though one may be dominant, the other recessive; one normal, the other defective.

This arrangement of the genes is one of the fundamental facts of genetics, which must be thoroughly grasped if the results of inheritance are to be understood. The arrangement of the genes in chromosomes; the fact that those thus together in the same chromosome usually go together into the same germ cell and the same individual; the fact that sometimes exchange of genes occurs between the two chromosomes of a pair; the fact that the frequency of exchange is greater the farther apart the genes are in the string of genes all these have most important consequences in inheritance and variation as generations pass. No one who neglects them can have an adequate understanding of the phenomena of heredity and variation, of similarity and diversity among individuals. These relations play a most important role in the further account of genetics given in the remainder of this book.

Can the genes be seen under the microscope? During the formation of germ cells, at the time when the crossing-over of genes occurs, the chromosomes are seen to be linear structures, with consecutive thickenings (chromomeres), the two thickenings of the conjugating pair of chromosomes side by side. While it is impossible to prove that these thickenings represent or contain the genes, it seems probable that they do. For their arrangement is exactly that which experimentation shows must be the arrangement of the genes.

MAPS OF THE CHROMOSOMES

Since the genes are thus known to be in serial order, maps can be made of them, provided there is a sufficiently detailed knowledge of the genetics of the organism in question. Such maps show the order of the genes in the chromosomes. In making such maps the guiding principle is the fact that two genes having together small exchange ratios are close together, and that the larger the exchange ratio between two genes the farther they are apart. The position of a gene on the map or in the chromosome is known as its locus. The locus of any gene is defined by its distance in certain units of length from one of the ends of the chromosome, the upper or 'left' end in the maps being that selected as the point of reference. Each unit is such a chromosome length as gives, for two genes at opposite ends of it, exchanges in one per cent of all cases (thus it is a trifle less than the distance in the X-chromosome from the gene yellow to the gene white, which exchange is about 1-5 per cent of all cases). To designate the locus of a gene, it is common to specify the chromosome in which it occurs (I, II, III or IV) and its distance in units from the upper end of the chromosome. Thus, in the X-chromosome, which is number I, the gene white is at the locus I, 1.5; the gene miniature at I, 36.1. The gene purple is at II, 54.5; the gene sepia at III, 26.0, and so on.

Such maps when carefully made can be depended on as showing correctly the order of the genes, and approximately their relative distances apart. The relative physical distances apart, as would be seen under the microscope, differ somewhat from those shown on the map, because the frequency of exchange is somewhat greater in some parts of the chromosome than in others. Also, the frequency of the breaks and exchanges in the chromosomes is known to be affected by a number of other conditions besides the distances separating the genes. Yet on the whole the relations shown by the map are sufficiently near to those actually occurring to make the maps of great value. They are extremely useful in helping to understand and predict the course of inheritance. From them can be determined the proportions of descendants that will show certain combinations of characteristics, when the combinations occurring in the ancestors are known, as will be brought out in later pages.

Maps of the chromosomes have been made for only a few organisms thus far, particularly for two or three species of Drosophila. Beginnings have been made for the chromosomes of maize, the pea plant, the sweet pea, the rat and mouse. For most organisms the experimental data are as yet too scanty to permit even the beginnings of chromosome maps to be made. This is notably the case for man. His 24 pairs of chromosomes together with the fact that experimental breeding cannot be carried on in man, will make the preparation of maps of his chromosomes a difficult if not impossible task.

PLACE AND FREQUENCY OF BREAKS IN THE CHROMOSOMES

Certain important points have been discovered as to the frequency of breaks, the number of breaks, and their location, in the exchanges that occur between the two members of a pair of chromosomes. The evidence for these cannot be given here, as it is detailed and voluminous, but the main facts may be summarized. It will be recalled that the breaks and exchanges occur in the germ cells at the time when the two chromosomes of a pair are side by side as long threads.

1. In the different cells of even the same individual, the chromosomes of any type, as X, or a particular autosome, show breaks in different numbers and locations.

2. In some of the cells the chromosome does not break, so that there is no exchange. This is, as a rule, the case in a rather large proportion of the different cells.
3. In some of the cells the chromosome breaks in but one place, so that, if we designate one of the chromosomes p_9 the other m, a single continuous piece of p including one end is transferred to m; a similar piece of m top.
4. In a small proportion of the cells the chromosome breaks in two places, so that a piece from the middle portion of p is transferred to m, and vice versa.
5. Very rarely the chromosomes break in three places, so that there is exchange of two separate pieces. More than three breaks are extremely infrequent.

The consequences in inheritance of the fact that a single chromosome contains many genes, and that exchange of genes occurs between the two members of a pair of chromosomes, are taken up in later chapters.

Chapter 9

The Genetic System as a Whole

Having examined the chief component parts of the genetic system, we proceed to an examination of the way it functions as a whole. This functioning and its consequences constitute what is usually called Heredity.

As we have seen, the genetic system consists mainly of the chromosomes with their genes: the X-chromosomes, the autosomes, and the Y-chromosome.

Why are the other parts of the germ cells not to be considered 'materials of heredity' or parts of the genetic system the cytoplasm as well as the chromosomes? It is known that the cytoplasm is of the very greatest importance in the development of the individual and in the production of its inherited characteristics. It is in the cytoplasm, through interaction with the genes, that the differentiations of the body arise.

The role of the cytoplasm, however, differs from that of the chromosomal materials in the following respects. Different germ cells differ effectively in their genes, and it is to these gene differences that the appearance of diverse characteristics in different individuals is due. But as a rule there is no evidence that the cytoplasm of different germ cells is effectively diverse in such a way as to produce different characteristics in different individuals. It is therefore the former the chromosomes with their genes that are commonly classified as the 'materials of heredity'more properly perhaps the 'materials of hereditary diversity'. The cytoplasm might correspondingly be called the 'materials of bodily differentiation.

Following the prevailing usage therefore, we may define the genetic system or material of heredity as follows:

The genetic system consists of those parts of the cells that are effectively diverse in different germ cells, these diversities causing different characteristics to appear in the different individuals produced.

In rare cases, in certain plants, it is found that the cytoplasm contains programmes bodies which may differ in different germ cells and so cause differences in the characteristics of the individuals developed from them. In such cases the cytoplasm, or at least the programmes bodies it contains, must be included as part of the genetic system. But in most organisms the genetic system consists of the chromosomes with their genes.

CONSTITUTION OF THE CHROMOSOMES THE GROUPS OF CHARACTERS LINKAGE

Each of the chromosomes (save in some cases the Y-chromosome) is composed, as we have seen, of many diverse genes. In the wild, unmodified individuals, in the fruit-fly, the genes are such as produce the 'normal' or 'wild-type' characteristics of the individuals. The genes of any chromosome become modified in the course of time, in the process known as mutation, which will be dealt with in a later chapter. The modified genes produce altered characteristics, most of which are recessive, while the normal characteristics, when mated to the modified characteristics, are dominant. A few of the modified characteristics, however, are dominant when mated with the normal or wildtype characters. The modified characteristics resulting from mutation are in most, if not all, cases defects, as compared with the normal characters. Thus the normal, dominant characters form a standard condition, of which the defective and recessive mutated characters are modifications.

The diverse genes of any chromosome are designated, so far as they have received names, by the name of one of the modified characters which it helps to produce. Thus the gene yellow, near the upper or left end of the X-chromosome of Drosophila, produces in its mutated condition the modified body programmes yellow; while in its unmodified condition it produces the normal grey programmes of the wild fly. In the maps of all the names apply to the modified or mutated condition of the gene. For every such modified gene there is a corresponding unmodified normal or wild-type gene. It is the custom to designate the wild-type gene from which any modified gene is derived by writing with the name or abbreviation for the modified gene the sign +. Thus yellow + or y + or + y signifies the normal wild-type gene that produces grey body programmes; white + or w + signifies the normal gene located at I, i .5, which produces red eyes in place of white ones.

The genes of any one of the chromosomes, as X, or a given autosome, become modified in many different ways, so that different examples of the same chromosome have different sets of genes. The diversity of genes in the chromosomes of different germ cells causes diversities in development, with resulting different characteristics in the individuals produced. The characteristics produced by the action of the genes of any chromosome follow in the individuals of successive generations the distribution to the offspring of the genes that produce them. Thus many characteristics follow the course of the X-chromosome; many follow each autosome; a few may follow the Y-chromosome.

Since a large group of characteristics follows each chromosome, the result is that characteristics pass from parent to offspring in groups. As an X-chromosome passes from the father into his daughters, all the genes, and all the characteristics that they produce, pass together into the daughters. Here, of course, some of the recessive characteristics may remain hidden in the presence of dominant genes in the second X- chromosome carried by the daughters. But all may reappear anew when in a later generation this X-chromosome is again by itself in a male. Except in so far as there is exchange or crossing-over with the other chromosome of the pair, all the characteristics that depend on the genes of a particular chromosome pass together into the individual that receives that chromosome, and this continues for generation after generation.

Thus, in Drosophila, suppose that a male has white eyes and yellow body, two characteristics connected with genes in the single X-chromosome. He is mated with a female that has red eyes and grey body, again owing to genes in the X-chromosome. In the daughters of these two, the two types of X-chromosomes are present together. These daughters produce offspring in a third generation. Some of their sons receive the X-chromosome that came from the original male; these will have white eyes and yellow body. Other sons receive the X-chromosome that came from the original female; these have red eyes and grey body. Only a very minute proportion of the offspring have the reverse combinations of characteristics white eyes with grey body, and red eyes with yellow body. The two characters originally together white and yellow appear to be linked, and the same is true of the other combination, red and grey. The original combinations do not readily separate.This linkage of characters in heredity was observed before it was known that characteristics depend on chromosomes. Many theories were invented to account for it, until it was discovered that it is due to the fact that the linked characters depend on genes that are in the same chromosome. All characteristics that depend on the same chromosome form what is called a linkage group. Since there are different numbers of chromosomes in different organisms, there are correspondingly different numbers also of linkage groups.

These linkage groups are of great importance in heredity. They reveal to us many things about the nature and action of the genetic system. We shall therefore examine these groups in a typical organism, selecting for this purpose the animal in which they are best known, namely, Drosophila melanogaster.

THE GROUPS OF LINKED CHARACTERS IN DROSOPHILA

The fruit-fly has, as we have seen, four pairs of chromosomes. On the genes in these four pairs of chromossomes depend several hundred known inherited characteristics. These many characteristics form four linked groups, according to their dependence on genes in the X-chromo- some, or in the chromosomes II, III or IV. It is to be noted that all of the characteristics that depend on genes in either one of a given pair of chromosomes constitute a single linkage group, since all those characteristics may become connected with the genes of a single one of the chromosomes. Thus the number of linkage groups is the same as the number of pairs of chromosomes.

In the course of breeding successive generations, linkage of charact-eristics shows itself in the fact that certain characteristics (dependent on genes in the same chromosome), which are present together in one parent, are present together also in the individual descendants. This is best seen on comparing grandchildren with grandparents. In this case one grandparent had the combination white-yellow, the other the combination red-grey. The same two combinations reappear in the grandchildren (seen particularly in the grandsons). In addition there are a very few 'cross-overs', having the new combinations white-grey in some individuals, red-yellow in others. But almost all the grandchildren get both the characteristics from the same gandparent. If the original grandparents have the other combinations, grey-white and yellow-red, then the majority of the grandchildren show these combinations.

Thus just what combination of characteristics the individual grandchildren have depends on what characteristics were united in the grandparents. But this is the case of course only for characteristics whose genes lie in the same chromosome. If two characteristics depend on genes that are in different pairs of chromosomes, then the combinations that were present in the grandparents are no more frequent in the descendants than are the reverse combinations.

It is in many respects most convenient to think of linkage in terms of the gametes. The combination of genes that are present in the gametes can often be determined by examination of the offspring produced. With relation to the gametes, linkage can be characterized as follows. When an individual (zygote) is formed by the union of two gametes, each bearing in one of its chromosomes a certain combination of linked genes, this same individual later produces gametes of which

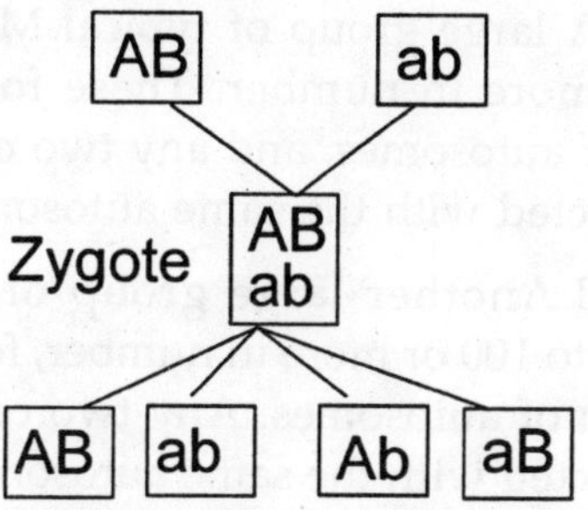

• Gametes • Gametes • Mary Few

Fig. Diagram of the results of linkage, with relation to the gametes. The two gametes above carry respectively the linked genes AB and ab. The zygote produced by their union carries both AB (in one chromosome) and ab (in the other). Then the majority of the gametes formed by this zygote carry again the linked genes AB or ab; only a minority cany Ab or aB.

the majority have these same combinations of linked genes. Often also there is a minority of gametes that have the genes combined in a new way, as a result of crossing-over.

In the fruit-fly, crossing-over occurs only in the females, not in the males. Thus if a male individual is formed by the union of two gametes carrying certain combinations of linked genes, then later such a male produces only gametes having the same combinations of linked genes. This fact is extremely convenient for determining what characteristics are linked in Drosophila.There are many other organisms in which crossingover is not limited to one sex. By breeding for successive generations and by the study of exchanges between the chromosomes, the following groups of linked characters have been found in the fruit-fly.

Group IA large group of sex-linked characters, more than 100 in number.These all follow in heredity the distribution of X-chromosomes, as described in earlier pages. Any two or more of these characters may be linked together and connected with the same X-chromosome.

Group IIA large group of typical Mendelian characters, 50 to 100 or more in number. These follow the method of distribution of autosomes, and any two or more of the group may be connected with the same autosome.

Group III.Another large group of typical Mendelian characters, 50 to 100 or more in number, following the method of distribution of autosomes. Any two or more of this group may be connected with the same autosome, but a character of Group III is not linked with any character of Group II.

Group IV.A small group of typical Mendelian characters, only 5 or 6 known.These are not linked with any of the characters of Groups I, II or III.All of the 300 or more characters known in this species fall into one or the other of these four groups. The nature of the characters found in the different groups is considered on later pages.

Thus here the number of linked groups is the same as the number of pairs of chromosomes. There are three large groups and one very small one; likewise there are three large pairs of chromosomes and one very small pair. It is natural to conclude, therefore, that the groups of linked characters correspond to the pairs of chromosomes, and that the characteristics in each group are connected with some particular pair. As it turns out, there is conclusive evidence that this is true. The evidence for each of the groups may be summarized as follows:

All the characters in Group I are sex-linked; that is, they follow exactly the distribution of the X-chromosomes. Any character in this group follows a particular X-chromosome and its descendants, wherever that chromosome goes. The characteristics in Group I are thus certainly dependent on genes in the X-chromosomes.

The small group of characters known as Group IV has been proved to be connected with the small chromosomes of pair IV. This has been demonstrated in the following manner. By non-disjunction of this fourth pair of chromosomes, some individuals are produced having but one small chromosome (IV). That this is the situation can be seen under the microscope, in properly prepared material.

These individuals with but one of the small chromosomes have the dominant normal or 'wild' characteristics. Such individuals are mated with individuals containing one of the recessive characters of Group IV. One of these recessive characters is 'bent wings'; another is 'eyeless' the individual having no eyes.

Half of the germ cells from the individuals that have but one of the fourth chromosomes are without this chromosome, while half of them have it. These two kinds each unite with germ cells carrying the recessive character 'bent', or 'eyeless'. There are thus produced individuals, half of which have but one of the fourth chromosomes, while the other half have two. In those which contain two, all the characteristics are of the dominant normal type; the recessive character bent or eyeless is not manifested. But in the individuals having but one fourth chromosome, the recessive character is manifested: the individual shows bent wings, or is eyeless. This demonstrates that the characteristics of Group IV are connected with the small chromosomes.

Thus the large linkage Groups II and III are left for the two large V-shaped chromosome pairs, II and III. These chromosomes can be broken into two or more pieces by subjecting them to radiations. When this is done, the characters of the corresponding linkage groups are no longer linked together.Thus in Group III the two genes scarlet and sooty are linked (completely in the male). But when the corresponding large chromosome (III) is broken into two pieces, scarlet and sooty are no longer linked; they may pass to different offspring.

They are evidently connected with the two separated pieces of chromosome III. In the same way the other large chromosome (II) may be broken, and this causes the characteristics of the linkage Group II to separate into two groups. Minute study shows that the chromosomes II and III differ a little, in size and form, so that with practice it becomes possible to distinguish one from the other under the microscope.

Thus it is proved that each of the four linkage groups is connected with one of the four chromosome pairs; the four groups correspond to the four chromosomes.

In a number of other organisms the number of linked groups of characters has been determined, and also the number of pairs of chromosomes. In every case in which this has been fully worked out, the number of linked groups is the same as the number of pairs of chromosomes. The numbers of groups and of pairs in certain organisms are as follows:

Drosophila melanogaster 4, Drosophila virilis 6, garden pea 8, sweet pea 8, maize10.

In man there are twenty-four pairs of chromosomes, including the X's. It is to be expected therefore that there will be found 24 groups of linked characters, one group sex-linked, the other 23 groups typical Mendelian, in their inheritance. But it will be a very long time before these 24 groups are identified.

FUNCTIONS OF DIFFERENT CHROMOSOMES IN RELATIONS TO CHARACTERISTICS

Have the different chromosomes different functions? Can we classify their functions in any simple or systematic way? Do certain chromosomes influence certain particular parts of the body, or certain particular functions, while others influence other parts or functions?

Such questions can be answered by an examination of the kinds of characteristics found in each of the linked groups; this will show what kinds of characteristics are influenced by each of the chromosomes. We give, therefore, the following lists of a number of typical characteristics connected with each chromosome in the fruit-fly. The characteristics listed are all modifications of the normal characteristics found in the typical wild fly. Hence for every characteristic here listed there exists also a corresponding unmodified normal characteristic. Each of the characters listed is the result of modification of a single gene.

List of some Typical Characters connected with the Genes of each of the Four Chromosomes of the Fruit-fly

In the lists, the part of the body, or the function, affected by the character is first indicated.This is followed by the names that have been applied to the different characteristics. Many of these are self-explanatory; of some others a brief characterization is given.

Group X-chromosome.Sex-linked inheritance.Eyecolour: coral, blood, eosin, cherry, apricot, buff, tinged, ivory, white (the foregoing are multiple alleles; that is, they are all diverse modifications of the same gene). Modifications of other genes: vermilion, ruby, carmine, garnet.

Eye form or structure: bar, facet, furrowed.

Body colours or markings: yellow, sable, lemon, green, chrome, tan, spot, dot, etc.

Wings: rudimentary, miniature, bifid, bow, notch, depressed, club, fused, cut, etc.

Legs: reduplicated.
Body structure: abnormal abdomen.
Bristles: forked, singed, tiny, bobbed, scute.

Reactions to light: tan (individuals showing this programmes characteristic do not fly toward the source of light, as do the normal individuals).

Life and health: at least 10 different lethal genes are known to be connected with the X-chromosome. Any one of these prevents the animal from developing, if no corresponding normal gene of the same pair is present. All, or practically all, the characteristics in the above list cause the organism showing them to be weaker and less resistant than the normal 'wild-type' flies.

As will be seen from the above list, the genes of the X-chromosome affect practically all parts of the body, and many of its functions.

Group II. Second chromosome (autosome). Mendelian inheritance.

Eye programmes or structure: purple, pinkish, cream, star, morula, etc.

Body programmes and pattern: olive, black, streak (thorax), comma, trefoil, speck, patched abdomen, etc.

Wings: truncate, balloon, vestigial, blistered, jaunty, curved, strap, apterous (and many others).

Bristles: fringed, minute, vortex.

Life and health: lethals (several).

Thus chromosome II affects many parts of the body.

Group III. Third chromosome (autosome).

Mendelian inheritance.

Eye programmes or structure: claret, cream III, deformed, kidney, maroon, pink, peach, rough, scarlet, sepia, etc.

Body programmes or pattern: band (thorax), ebony, sooty, trident, etc.

Body structure: bithorax, dwarf, giant.

Wings: beaded, spread, tilt, truncated, etc.

Bristles: doubled, hairless, hairy, minute, spineless, two-bristle, vortex III, etc.

Life and health: several lethals.

Thus chromosome III also affects many parts and functions.

Group IV. Fourth chromosome (autosome).

Mendelian inheritance.

Eyes: eyeless.

Wings: bent; cubitus interruptus.

Bristles:shaven.

Few genes known, but affecting diverse parts of the body.

What does the foregoing list of characters affected by the different chromosomes show as to the nature and functions of the chromosomes?

1. All the large chromosomes (Nos. I, II, and III) affect all parts of the body: many different functions, health, length of life, and the like.

   The small pair (IV), though affecting few characters, still affects diverse parts (eyes, wings, bristles).

2. Thus particular chromosomes are not limited in their action to any particular part of the body or to any particular function. Each chromosome affects many parts and functions.

3. It appears clear that each chromosome enters into the process of development, affecting many processes, and doubtless influencing the entire organism. The method of action of the X-chromosome has been treated in a previous chapter; as there seen, it alters the chemical processes, thus affecting the development of all parts of the body.The other chromosomes doubtless act in a somewhat similar manner.

4. It is not clear from our present knowledge that the genes are arranged in the chromosomes in any systematic way, with relation to their functions, or to the parts of the body that they chiefly influence. A gene principally affecting eye programmes is close to one principally affecting body structure or body

programmes or structure of the wings, and so on. Genes affecting eye programmes occur in many different parts of the genetic system. In a similar way, genes that affect the structure of the wing or of the eye, or that affect the programmes of the body, are scattered throughout the four chromosomes.

Thus if there is any functional system in the arrangement of the genes in the chromosomes, it has not yet been discovered; in the present state of our knowledge it appears as if there were none.

By 'characteristics' or 'characters' are meant any features, physical, chemical, physiological, or structural, in which organisms may differ. The materials presented in the chapters preceding this one have brought out, explicitly or implicitly, many of the relations between genes and characteristics. Here these relations, and others not yet set forth, are brought together in systematic order.

1. SPECIAL ACTION OF A GENE

Particular genes affect mainly, though not exclusively, some particular part of the body, or some particular function.

Thus, in Drosophila there is in the X-chromosome, at the locus I, 1.5, a certain gene which in its normal condition takes part in producing the red eye programmes. This gene has become modified in different cases in a considerable number of different ways, and any modification alters the programmes of the eye. The different modifications give the many different alleles listed earlier among the eye-programmes characters of linkage Group I. Other genes have their chief effect on the body programmes or on the wings or on some other part. Particular genes in man affect the blood type of the individual; others affect the eye, the body form, and so on. Each gene as a rule has its main function in connection with the development of some particular part.

2. Yet a given gene often produces effects on several parts or functions. For modification of a single gene may cause a change in the characteristics of a number of different parts. Some examples of this may be given.

   The character bent wings, belonging to linkage Group IV of Drosophila, is due to the modification of a single gene in chromosome IV. It causes the wings to be bent dorsally at their base, and also it causes the legs to be shorter than normal, owing to the abnormal shortness and thickness of the meta- tarsal joint.

   In animals showing the character vestigial wings (due to modification of a gene in chromosome II), the following peculiarities appear: the wings are degenerate, only their bases being present, and these are held at right angles to the body. The balancers are likewise degenerate. Two rear bristles on the scutellum are farther apart than usual, and they stand up straight instead of inclining backward. The flies hatch about two days later than in the normal wild flies. They are not quite so vigorous as the normals.

   As noted in the preceding chapter in connection with linkage Group I, modification of a certain gene in the X-chromo- some causes;

   a. The body programmes to be tan instead of grey,
   b. Causes the individuals to lose the usual positive reaction to light.

   In any detailed account of the effects of particular genes, it will usually be found that each gene has thus several different effects, often on diverse parts of the body. The effect on which the name given to the gene is based is commonly merely that effect that is most conspicuous to the eye.

3. General or Constitutional Effect of a Gene

In addition to the special effects on certain structures or functions, modification of any gene has also a general effect on the bodily constitution. In Drosophila the 'wild-type' flies, with unmodified genes, are vigorous and long-lived. Practically all modifications of the normal genes make the animals less vigorous, so that individuals showing any of the characteristics listed in the preceding chapter are weaker than the normal or 'wild- type* individuals, and have a higher rate of mortality. Thus practically all the modifications of normal genes are defects, a fact to which we return in a later chapter.

In view of the facts set forth in preceding paragraphs it has been said with much probability that 'every gene affects the development of the entire body.

4. Thus apparently each gene supplies material which in the process of development enters into reaction with materials from all the other genes; it thus modifies many or all parts of the body, but affects certain parts more conspicuously than others.

5. As before set forth, a particular single gene may become modified in many different ways, in different individuals or different chromosomes, each modification causing a changed characteristic. The classical example of this is the series of different eye colours (multiple alleles) found in different individuals of Drosophila, each programmes being due to a different modification of the gene located at the point I, 1.5 in the X-chromosome. This gene in its normal condition produces (by interaction with other eye-programmes genes) the normal red programmes. The various modifications of the gene give rise to the colours before mentioned: coral, blood, eosin, cherry, apricot, buff, tinged, ivory, white.

Many such series of multiple alleles are known in the fruit- fly and in other organisms. In the fruit-fly the following illustrate their nature:

Scute: The gene scute is close to the upper (or left) end of the X-chromosome. It exists in 20 to 30 different modifications, each of which causes a slight change in the distribution of the bristles on the body.

Cut: A considerable number of different alleles are known of the cut gene, located at I, 20.0. All of them cause changes in the outline of the wings, as if bits of the wing were cut off.

Vestigial: The gene vestigial at II, 67, which reduces the size and changes the form of the wing, has two other alleles, antlered and strap> giving wings of other forms and sizes.

Garnet: At the locus I, 44.4 are several alleles which change the programmes of the eye in diverse ways. Some of these are known as garnet, salmon, garnet 2, garnet 3.

Pink: The eye colours, pink, peach, pink 3, pink 4, pink 5, are a series of alleles located at III, 48.0.

Series of alleles are known also in other organisms. Thus in the rodents (rats, guineapigs, rabbits), series of different alleles are known, yielding different colours of the hairy coat. In maize there are several allelic series modifying the programmes of parts of the plant. Presumably any gene may, by successive modifications, produce a series of alleles.

As mentioned in an earlier chapter, all the members of an allelic series affect the same character or part of the body, yielding different modifications of that character. Further, they show certain peculiarities of inheritance as compared with different genes that are not alleles. Since they are all modifications of the same gene, only one of the alleles can be present in a single chromosome, and for this reason only one of them can be carried by a single gamete (except in cases of non-disjunction). Further, only two members of a given allelic series can be present in any diploid individual or zygote.

There are thus two types of diversities among genes,

a. On the one hand, there are the different genes occurring at different loci in the chromosomes, the diverse genes, shown on the maps affecting many diverse characteristics.
b. On the other hand, there are these different modifications of a single one of these genes the multiple alleles.

6. Any given single character, inherited in the sex-linked or the autosomal manner, is a product of the interaction of many genes, located in different parts of the genetic system.

   As an example of this, the programmes of the eye in Drosophila may be taken. As shown in the preceding chapter, the programmes of the eye depends on many genes scattered through the different chromosomes. The number of different genes directly affecting the programmes of the eye is indeed still greater than this, and in addition there are many other genes that affect eye programmes indirectly, through the fact that they affect the structure of the eye: for example, the gene 'eyeless* in the fourth chromosome pair. If this gene is modified in a certain way, no eyes are developed, consequently no programmes.

   Similarly, many different genes affect the form and structure of the wings, the venation of the wings, the structure of the eye, the programmes of the body, the form of the body, the structure and distribution of the bristles. Indeed, this is presumably the situation with relation to any characteristic whatever; every characteristic is influenced by many genes.

7. Altering any one of the numerous genes that cooperate to produce a given characteristic alters that characteristic. Any characteristic may, therefore, be changed by altering any one or more of many different genes.

   Thus, consider what must be the situation among the genes to yield the normal red eye programmes in Drosophila. To produce this result all of the 19 (and more) genes that affect eye programmes must be present in the usual or normal condition. If any one of them is changed (the rest remaining unchanged), the programmes of the eye is altered.

Changing in a certain way a gene at I, 1.5 all other genes remaining normal, the eye programmes is changed to white. If we change the gene at I, 33 (the rest remaining normal), the programmes becomes vermilion. If we change the gene at II, 54.5 (the rest remaining normal), the programmes of the eye becomes purple. Thus one could go over the entire 19 genes; changing one after the other; each one changed alters the eye programmes.

Again, consider what must be the situation in the genetic system to yield one of the recessive colours: for example, vermilion eye programmes. To produce this programmes, 18 of the genes must be in their usual normal condition, while a single one, at the point I, 33, must be altered. If, in addition to this one, another is altered (for example that which by itself yields purple, at II, 54.5), a different programmes is produced. Altering one gene changes the programmes in a certain way; altering two or more changes the programmes in a still different way. Similar relations would be found with respect to most or all characteristics.

8. Multiple Factors

The different genes on which a given character depends may have qualitatively different effects; such is the case with the different genes that affect the eye programmes in the fruit-fly. In other cases the different genes affect the characteristic only quantitatively. Thus dimensions often depend on many different genes. Modifying one of these or substituting another for it changes the dimension; modifying an additional one changes it still more, and so on. In such cases the character is commonly said to depend on 'multiple factors'.

9. 'Duplicate Factors

In some cases a character may depend on the presence of two (or more) genes of a particular type in such a way that both of these genes must be present in order that the character shall be produced. The character is then said to depend on duplicate genes or factors. In the plant known as the shepherd's purse, the common type of pod is triangular, but some plants have an elongated oval pod. This oval pod occurs only when two particular recessive genes are present in the homozygous condition. Thus if the two dominant genes which produce the triangular pod are A and B, while the two recessives that give the oval pod are a and b, the oval pod occurs only in the plants having the genes aabb, while all the other combinations give the triangular pod. Similar cases of duplicate genes are known in a number of organisms.

10. 'Modifying Factors'

a. In some cases the effect of some of the genes is merely to modify in detail a characteristic that is due primarily to another gene. Such are known as modifying genes or factors. Thus, in Drosophila, a certain modified gene at I, 1.5 causes the eye to assume the pinkish programmes known as eosin, but the depth of this eosin programmes depends upon certain other genes present in other parts of the genetic system. One of these, known as cream b, is located at II, 22.5; another, cream III, at III, 36.5; and several other such genes modifying the eosin programmes are known. If these genes are present in the normal or unmodified condition they do not affect the eosin programmes; when modified in certain ways they do. Again, in the rat, the presence of a certain gene causes the fore part of the body to be coloured brown, forming what is called a hood) the rest of the body being white. The extent of this hood depends on the presence in the individual of a number of other genes; some of them increase its extent, others decrease it. Such modifying genes or 'modifying factors' play a role in many characteristics. They may produce no visible effect except in the presence of a particular character that they modify. Thus the modifiers of eosin eye programmes produce no visible effect in an individual with red eyes.

b. The same phaenotypic character that is, the same characteristic in outward appearance may be due in different cases to the alteration of diverse genes, in different parts of the genetic system. This is illustrated in numerous cases in Drosophila.

Thus, modification of a certain gene at II, 54.5 causes the programmes of the eye to change from the usual red to a darkish purple; the modified gene giving this programmes is known as purple. But in other cases what is outwardly the same programmes is found to be produced by change in other genes. One of these is at I, 44.4; to distinguish it from the one just mentioned it is called garnet. A third gene that produces practically the same eye programmes is found at III, 49.7; it is called maroon.

Similarly the eye programmes 'salmon', resulting from a modified gene at I (again at 44.4), is indistinguishable from the eye programmes 'rose', resulting from a gene modification at III, 48.0.

Other cases of this sort could be given. In such cases the same phaenotypic character will be found to be inherited in different ways in different cases, depending on the location of the modified gene to which the character is due. If the gene is in X, the inheritance will show itself to be of the sex-linked type; if in II, it will be autosomal and of the linkage Group II; if in III, it will be autosomal and of the linkage Group III.

THE ACTION OF GENES IN DEVELOPMENT

In view of what has just been set forth, how are we to conceive the operation of the genes in producing the characteristics? How do they operate in producing an adult from the fertilized egg? And how do different sets of genes produce individuals having different characteristics?

On the intimate details of the operation of the genes little is known. Some of the main features in their method of action were taken up earlier in dealing with the relation of chromosomes to sex, though only the action of large groups, forming chromosomes, was there dealt with. From the relations brought out above, and in preceding chapters, taken in connection with the facts of embryological development, it is possible to form a conception of some of the main features in the action of genes in building up the body. These are summarized in the following.

The genes form a set of many diverse materials, differing from each other chemically and physically and having different chemical and physiological activities. They exist at the beginning of the individual's development in the nucleus of the fertilized egg, while surrounding them is a mass of cytoplasm.

The genes interact with the cytoplasm, changing it, producing new materials that are incorporated into the cytoplasm. These varied cytoplasmic products of the different genes interact with each other, producing again new types of material.

Some of the gross features of the interaction of genes and cytoplasm were described earlier. As there shown, the chromosomes (groups of genes) visibly take up material from the cytoplasm, alter it, and give it off again into the cytoplasm.

Though the genes thus interact with the cytoplasm, modifying it, they are not themselves used up in this process so as to lose their essential character. They supply material which combines with cytoplasmic material. But there is left an unchanged part of each gene, and this by division produces additional genes of the same type and action.

As the cells divide, each chromosome divides, and each of its contained genes divides. One of the two halves of each gene goes after division to each of two cells produced. Thus each new cell produced contains the full set of genes, all those that were present in the fertilized egg.

In most organisms it appears probable that this continues throughout at least all the early stages of development, if not to the end of development, so that all the cells contain all of the genes. For in the young embryo, after many hundred cells have been produced, it can be shown that any cell can produce many different parts of the organism. And in many organisms even in late stages there are cells that can produce any part, or the whole organism, as is shown by the facts of regeneration. Such cells must contain the entire set of genes.

In very late stages of development it is possible that in highly differentiated tissue cells some or all of the genes have become lost or modified, though there is no clear proof of this. Further, there are a few organisms of which it is to be observed that the cells that are to produce the differentiated body of the animal lose parts of their chromosomes, and so presumably some of the genes. The chromosomes in these cells are seen to break up, and portions of them are dissolved and absorbed by the cytoplasm while in the germ cells the chromosomes remain entire. This is known to occur in threadworms (Nematodes) and in certain insects. The relation of these processes to the way bodily development occurs is quite unknown.

But on the whole the evidence is strong that in most organisms the production of different tissues and organs by different cells is not due to their containing different genes, for cells that certainly contain all the genes produce many different parts in many different cases. The genes act differently in the different cells, depending on the relation of the different cells to each other, so that the different cells, though containing the same genes, produce different tissues and organs.

The diversity of the different parts of the body seemingly arises mainly, if not entirely, through the fact that different cytoplasmic products of the genes are separated into the different cells. Thus the cells become different in their cytoplasm. And this different cytoplasmic constitution of the different cells interacts differently in each case with the full set of genes that the nucleus contains, so giving anew different products. By a continuation of such action the different tissues and organs are produced. In each, the cells contain the same sets of genes, but differ in their cytoplasm. The differentiations of the body therefore lie mainly in the cytoplasm, though the diversities in the cytoplasm are brought into existence through the action of the genes.

Yet it is true that differences may arise within the body of a single individual through a change in the genes of some of the cells. As before seen if in a female Drosophila one of the cells loses one of the X-chromosomes, that part of the body derived from this cell shows the characteristics of the male, while the rest is female. If the X that is lost carries dominant genes, while the remaining X has the corresponding recessive genes, then, in that part of the body whose cells have lost one X-chromosome, the recessive characters are manifested, while the remainder of the body is dominant. In this way are produced mosaic organisms, showing diverse hereditary characteristics in different parts of the body. Such mosaic diversities, however, appear to be of a very different type from the differentiation of the body into tissues and organs.^Whether any differentiations in ordinary development are thus produced through loss or modifications of certain genes must be considered as yet uncertain, though it is entirely clear that most of them are not so produced.

But diversities between individuals, as we have seen, commonly result from differences in the genes that they contain. If two fertilized eggs begin with different sets of genes, the processes of changing the cytoplasm, of differentiating it into tissues and organs, differ in the two cases, so that individuals with diverse characteristics are produced. Very marked differences in characteristics may be produced even when the two sets of genes differ in but a single gene or pair of genes. In Drosophila, as before seen, if two eggs differ in a single pair of genes in chromosome IV, one of them produces individuals without eyes, while the other has normal eyes. If two individuals differ in an entire chromosome (X), one becomes a male, the other a female.

Such differences are due to the diverse chemical and physical action of the different genes. They produce differences among different cells, in the way set forth in earlier paragraphs. Also, they may produce diverse chemicals (hormones) which circulate through the body, thus causing changes in the development of parts with which they come in contact.

Different chemical and physiological effects, different development, may be produced by difference in balance among the genes. Two cells may contain the same kinds of materials (genes) throughout, but one of these contains certain materials in two doses (a pair of genes), while the other contains but one dose (an unpaired gene). In consequence the two cells differ greatly in development. In this way difference in sex is produced, with its manifold attendant differences in structures and functions. In many cases two genes of a certain kind produce a different effect from a single one, though in other cases they do not. That is, two centres of a certain kind of action may give different results from a single centre. This principle of the effects of difference in balance is presumably of great importance, though as yet not a very great deal is known as to its operation.

It appears possible that the operation of some or all of the genes begins in early stages of development and continues throughout. It is known that some genes produce their distinctive effects at the beginning of development, in the egg itself. Thus in the silkworm a difference of one gene is known to make a difference in the programmes of the egg. Two races differ in egg programmes, one having slate-coloured eggs, the other brown eggs. Crossing the two races shows that the difference in programmes is due to a difference in a single gene, for the descendants show 'unit difference' inheritance. If, by crossing, the special gene from the slate-coloured race is brought into the eggs of the brown race, in the next generation the eggs are slate-coloured instead of brown, so that the 'slate' gene is active in the formation of the egg.

Other genes are known to produce their distinctive effects somewhat later, but still rather early in development. Thus in mice, if one of the genes of a certain pair is modified in a certain way, this causes the hair of the individual to be yellow. If both the genes of this pair are so modified as to produce the yellow programmes, the individual develops for but a short time, then dies while still an embryo. It is clear that these genes are active in the young embryo. Again, in plants, certain genes are required for the production of chlorophyll. If one of these becomes defective, chlorophyll is not produced. The young plant grows so long as the stored food in the seed is available, but, having no chlorophyll, it cannot make use of the sun's rays for the elaboration of nutrition; it therefore dies as a young seedling. Examples of the operation of genes during early development are abundant.

Some genes are required for laying the foundation of certain parts or organs; if these genes are altered, no such organ is produced. Thus in the fruit-fly a certain gene in chromosome IV is necessary, in its normal condition, for producing eyes. If that gene becomes defective, the eyes are completely lacking, although all the other genes that cooperate to produce the eye may be present in normal condition. This particular gene is required for the very foundations of the eye. Similarly, in vertebrates a certain gene is required for the production of the pigment which produces the colours of hair and skin. Many different genes cooperate to produce this pigment; most of them modify the type of pigment produced. But if this particular one becomes defective in a certain way, no pigment is produced; the animal is an albino.

A somewhat similar situation appears to be at the basis of some cases of feeblemindedness in man. In such cases the brain is not properly constituted; it does not function well. Study of inheritance shows that in some cases, at least, feeblemindedness is due to a defect in a single gene; there is 'unit difference inheritance. This single gene plays a necessary part in producing an efficient brain; when the gene is defective the brain is fundamentally imperfect.

The special and distinctive functions of some genes are not known to come into operation until late stages of development. Such, for example, are those that produce in Drosophila the distinctive pigments of the eye. Their visible effect does not appear until the eye is formed; although of course this action may have begun early in development.

In many such cases in which the conspicuous effect of a gene appears at a relatively late stage in development there are less conspicuous effects that indicate that the gene may be in operation throughout development. Thus individuals with white eyes, or with rudimentary or vestigial wings, are less vigorous than those with normal eyes or wings; they are less resistant and have shorter lives, and this is true for most of the types having recessive characters. Though the conspicuous special effect of the changed gene is late in appearing, the effect on the general constitution indicates that there may be an effect throughout development.

That this is indeed the case is indicated strongly by certain other facts. Sometimes a gene at a certain locus in the chromosome becomes entirely inoperative; it has either dropped out of the chromosome, or it has totally lost its action. Such cases are commonly known as deficiencies. Deficiencies are producible by the action of radiation on the chromosomes, so that large numbers of them have been studied. A deficiency is detected in the following manner. A chromosome suspected of being deficient for a certain gene at a certain locus is brought into an individual in which the other member of the pair to which it belongs carries a recessive gene at that locus. If the chromosome is deficient in the suspected gene, the recessive gene is manifested, since there is present no dominant gene of that pair. Thus if an X-chro- mosome carrying white eye is brought into a female whose other X is deficient for the gene at the white eye locus (the locus 1.5), the female has white eyes.

In the case of almost any gene, if two chromosomes that are deficient for the same single gene are brought into the same fertilized egg, so that the egg lacks that gene entirely, the egg does not fully develop. In most cases it is not known at just what stage development ceases. But the fact shows that, in addition to its special effect on a particular characteristic, the gene is necessary if development is to continue at all beyond a certain stage. By ingenious methods Demerec has recently proved that if only a few of the cells of a developing embryo are deficient for a certain gene, those few cells fail to develop, though the rest of the body 'does develop. This was demonstrated for deficiencies of many different genes in the X-chro- mosome of Drosophila. In only one of the many cases tested was deficiency for a certain part found not to be lethal, though even in this case it caused weakness. In the case of the white eye gene at I, 1.5, and in many other loci, deficiency prevented development.

Altogether, it is clear that genes are active throughout development, some beginning early, others seemingly later, and that almost all genes are necessary in order that full development shall occur. Changing one or more genes may cause from the beginning a change in the chemistry and physiology of the organism, altering its course of development and so giving rise to changed characteristics.

WHAT KINDS OF CHARACTERISTICS ARE AFFECTED BY GENES?

The question as to what kinds of characteristics are affected by genes has been much discussed. Some have supposed that only a few characteristics depend on genes: only certain superficial matters, such as colours, and the forms of external features. Particularly with relation to man has there been dispute on this matter.

But on this question, positive knowledge is available, both for man and for other organisms. It is possible to answer fully the question: What kinds of characteristics are affected by genes?

The question can mean only: What kinds of characteristics can be altered by changing the genes? What kinds of differences in characteristics result from differences in genes?

There are several ways of answering these questions. The most direct is by the crossing of individuals having different genes. This results in substituting one type of gene for another, with consequent differences in characteristics. If the diversities in characteristics show Mendelian or sex-linked inheritance, this demonstrates, as we have seen, that they are due to differences in genes.

By this test, in many organisms (for example Drosophila), the following kinds of characters have been found to be affected by genes:

1. Structures and forms of all parts (head, eyes, thorax, abdomen, wings, legs, internal organs).
2. Colours and patterns of all parts. These depend on chemical processes, hence:
3. Chemical processes of many kinds are affected by genes.
4. Physiological conditions: health, vigour, life and death.
5. Power of development: certain lethal genes stop development at a certain stage. A defective gene in the fourth chromosome of Drosophila prevents the development of the eye. Certain genes cause the body to grow to a larger size than usual. Many examples of the effects of particular genes on development have been given in earlier pages.
6. Behaviour. Reactions to light in Drosophila are changed by alteration of a certain gene ('tan').
7. Sex. Changing the genes changes the sex; this involves changes in all types of characteristics: in structure, pattern, programmes, physiology, behaviour.

Thus, in sum, it is clear that in animals, characteristics of all kinds, without exception, are affected by genes. There are no characteristics that are exempt from influence by the genes.

Any characteristic can be altered by suitable alteration of genes. This is not surprising when it is considered that the genes are among the materials from which organisms are made, and that altering them alters the chemistry of the developmental processes.

What is the situation in Man? What characteristics are affected by genes in Man?

In man we have to deal with certain important types of characteristics that are hardly to be studied in other organisms. These are what we call mental characteristics: emotions, temperament, skill, character, intelligence, reasoning power, and the like. There are great differences among human beings in their mental characteristics, and in their capabilities. Are these things affected by genes? Often it is argued that they are not. Some have held that practically all differences in mind and character are due to environment, to education, influence of associates, tradition, state of civilization, and the like.

On this matter there is positive knowledge which completely removes the matter from doubt. There are two methods by which it can be determined in man what kinds of characteristics depend on genes.

1. The first method is by observing what characteristics show Mendelian or sex-linked inheritance. This method is not entirely satisfactory in man, since;
   a. It is not possible to carry on experimental breeding in man,
   b. This method can hardly be applied to the important mental characteristics. Yet by this method many characteristics of man have been shown to depend on genes. Among the characteristics showing sex-linked or Mendelian inheritance, and so depending on genes, are the following in man:
   i. Programmes of hair and eyes; form of many facial features; structure of the hands and feet; form of the body, whether stout or slender;

chemical constitution of the blood; efficiency of the brain (whether the individual is normal or feebleminded); condition of the eyes and sense of sight (colourblindness, short-sight, night-blindness and other conditions); sense of hearing (deafness, etc.); sex, and many other characteristics.

2. A second method gives much fuller and more detailed knowledge of what characters are affected by genes in man. This is by the study of identical twins, as compared with other individuals. This is a most important method, and it will therefore be discussed in some detail. It depends on the folio wing facts:

   a. In man, most individuals are derived from separate germ cells, and therefore have diverse sets of genes. For in forming germ cells, as before described, reduction of chromosomes and genes occurs, so that practically every germ cell differs in its gene combination from every other. When by fertilization two different germ cells unite, the diversity of gene combination is increased still more. The result is that no two individuals formed in this way have the same combination of genes.

   b. But rarely there occur cases in which a single fertilized egg produces two individuals (twins). Such individuals are formed by division of all the genes, so that the two twins contain the same gene combination. These are the twins known as monozygotic or identical twins. In another type of twins, the two are derived from separate eggs; such dizygotic twins have diverse gene combinations.

By comparing the similarities and differences of identical twins with those in other individuals, and particularly with those of twins of the other type, it is possible to find out what characteristics are affected;

1. By having all the genes alike,
2. By having different combinations of genes.

'Identical twins', or offspring all having the same genes, occur in many organisms. In lower animals (Protozoa, plants, Hydra, worms, etc.), a single individual often divides into two or more, by Vegetative* reproduction. All these individuals have the same sets of genes, since in vegetative reproduction every chromosome and every gene divides. Such individuals are very exactly alike. They have been studied a great deal. In such a Protozoan as Paramecium, a great number of individuals are produced in this way, all closely alike. But when conjugation takes place, this gives new gene combinations, and the individuals after they have conjugated are very different.

In one of the vertebrates, the armadillo, identical twins are formed in essentially the same way that they are in man, but are formed regularly. A single fertilized egg, after developing some distance, divides into four, so that four 'identical twins', or quadruplets, are produced at each birth. The four are extraordinarily alike even in fine details. But twins belonging to different sets (and so having different gene combinations) are often very different.

IDENTICAL TWINS IN MAN

Identical twins are produced in man by the division of a single egg into two, so that they may be called monozygotic or 'one-egg' twins. There occur also dizygotic or 'two-egg' twins, produced from two separately fertilized eggs; these are often called fraternal twins. Methods have been discovered by which it is possible to determine to which class any pair of twins belongs.

Identical twins in man are of enormous interest; they supply exactly what is required for discovering what characteristics depend on genes. They are two individuals that are made throughout from the same set of materials, from the same set of genes. Every single gene of one has its exact duplicate in the other. The two therefore have the same heredity; the differences between them are due to environment alone. They show how like or how diverse will be two human beings that do not differ in their genes, in their heredity, although they may differ in the conditions of their lives, in the experiences they undergo. Identical twins thus form one of the most interesting and illuminating of the phenomena that nature presents to us. They yield the key to that ancient question, the relative role of nature and nurture: of heredity and environment.

The existence of the two-egg twins, with their diverse sets of genes, for comparison with the one-egg twins, having identical sets of genes, completes what is required for discovering just what resemblances and differences in human beings are due to identity and diversity of genes. The conditions to which the two members of a pair of twins are subjected are as much alike in the two-egg twins as in the one-egg twins. Therefore we can be certain that any differences between the two types of twins are due to differences in the genes, not to diversities of environment.

Comparison of one-egg twins (having the same genes), with two-egg twins (having different genes): Identical twins are extraordinarily alike in all physical respects. They are always of the same sex. Their faces and Figs are so much alike that it is difficult to distinguish one from the other. They are alike in eye programmes, hair programmes, skin programmes. Their facial features are of the same form and relative size. Their teeth are alike, even to the irregularities. Their stature and weight are very similar. Their finger prints are as much alike as those of the two hands of the same individual.

In all these respects the two-egg twins are often very different. They are no more alike than any other two members of the same family.

Identical twins are also alike in weaknesses, in tendencies to certain diseases. If one has tuberculosis or other constitutional disease at a certain time of life, the other one usually has it also at about the same time, even though the twins live apart. But two-egg twins do not show these resemblances.

All this demonstrates that all these physical characteristics in man depend on genes. They are alike in two individuals whose genes are alike, diverse in two individuals whose genes are diverse.

One-egg twins at times show certain physical differences, resulting either from certain features of the process of splitting the original egg from which the two were produced, or from the fact that one of the twins has been injured by accident or disease. From the former cause arise the facts that (1) sometimes one of the twins is right-handed, the other left-handed, and (2) one is sometimes more vigorous than the other. If the splitting of the original single egg occurs very early, before the right and left sides have become diverse, the two twins are alike in 'handedness' and in vigour; if the splitting occurs later they may be diverse in these respects. The effects of diversity of environment on twins we take up in a later chapter.

MENTAL CHARACTERISTICS IN TWINS

In a number of cases the two twins from a single egg have been separated while very young ýÿbefore the age of one or two years, and have lived apart afterward. In such cases the genes of the two individuals are alike, while their environments have been different. What is the situation as to their mental characteristics?

Several cases of this sort have been very carefully studied by Muller and by Newman. We shall have occasion to look at the results of this study in some detail later. Here we may summarize briefly from our present point of view what was discovered. One-egg twins that have lived apart show the same close resemblance in physical characterization that is found in such twins as have lived together. They show also resemblance in mental characteristics: in emotions, temperament, and in 'intelligence' as measured by psychological tests. In certain cases, there are found some dissimilarities in mental and emotional characteristics, depending on differences in education and experience in the separated twins.

Thus it is clear that (1) the genes have great influence on mental characteristics, emotions, temperament, and the like, but that (2) these things may be modified by different education and experience.

An illuminating study of the mental and moral characteristics of twins was made by Johannes Lange. He found in the prisons of Germany, or in the prison records, thirty persons, each of whom was a member of a pair of twins, the two members in each case being of the same sex. He traced and examined the other twin of each pair. Thus one twin was chosen because he was a criminal; the other twin was taken as he chanced to be. As it turned out, thirteen of the pairs were single-egg twins, while the other seventeen were two-egg twins.

Of the thirteen pairs of one-egg twins, one of which was known in each case to be a criminal, it turned out, upon investigation, that the other also had a criminal record in 10 out of the 13 cases. It is clear that if one such twin comes in conflict with the law, the other one, identical in original nature, has small chance of escaping that fate. But with the two-egg twins, made up of diverse original materials, the situation was different. Their environments were as similar as were those of the one-egg twins. But of the 17 pairs of this kind, in only three cases was the other member of the pair found to have a criminal record. In the other 14 pairs the fate of the two was quite diverse. One was a criminal in each case; the other was not.

Thus it is clear that everything that helps to determine whether a man shall be a criminal or a good citizen is deeply affected by the materials of which the individual is made, by his 'heredity'. Mentality, morality, conduct, career, fate, are affected by the hereditary materials with which the individual starts life.

The truth of this conclusion came out still more strongly when the detailed records of the twins were studied. In the case of the single-egg twins, the two members conducted themselves as duplicate personalities might be expected to do. One such twin was a burglar; on looking up his brother's record, he was found to be a burglar too. In another pair, one twin was a high-powered financial swindler, fraudulently collecting large sums from his dupes. And so also was the other twin of that pair. Besides the pair that were burglars, and the pair that were swindlers, another pair of one-egg twins both 'committed puerile offences against the property laws'. Another consisted, in the words of Lange, of 'guttersnipes, but good fellows at heart', though they 'cannot stand alcohol; it develops frenzy and draws the knives from their pockets'. Another pair described 'have too little sense and will power', while in still another, both 'are lacking in all human feeling except for their own unpleasant selves'. And so goes the account throughout the entire list; the similarity in conduct is astonishing. Lange says, 'In all these pairs the type of crime is identical, the criminal careers begin at about the same age, and the behaviour of both members in court and in prison corresponds absolutely.'

But in the two-egg twins, that started life with different parental materials, with different heredity, the situation was different. As we have seen, 14 out of the 17 had totally different careers, one being a criminal, the other not. And there was no such detailed parallelism in the records of the two as was found in the one-egg twins. As before remarked, the similarity of environment was as great in the two-egg twins as it was in those that came from one egg. But this similarity of environment did not result in identity of fate or in detailed parallelism in careers. It was the identity of original nature, of the materials out of which they are made, that gave the same careers and the same fate to the one-egg twins. They were almost as much alike in their behaviour, their mentality, their career and their fate, as they were in their physical featuresýÿin which, like other one-egg twins, they were almost indistinguishable.

It is certain, therefore, that all the things that affect character and conduct are deeply influenced by the hereditary materials. There is no characteristic or quality of human beings that is exempted from its influence. This conclusion is confirmed by all the many studies that have been made on the two types of twins. And it is in harmony too with all that we know of the science of genetics. Experimental work on the breeding of other organisms shows, as we have seen, that they have no characteristics of any kind that are not affected by the genes that they receive at the beginning of their lives.

Are characteristics influenced also by other conditions, by the environment? The fact that all are affected by genes does not at all imply that they may not be affected in other ways.

Chapter 10

Rules and Rations of Inheritance

In earlier chapters we have seen that the dependence of characteristics on genes located in the different chromosomes results in three diverse types of inheritance. Characteristics follow in descent from earlier to later generations the distribution of the particular chromosomes in which are the genes on which the characters depend. Since there are three different types of distribution of the chromosomes, there result three types of inheritance. These are:

1. Sex-linked inheritance, dependent on genes in the X-chromosome;
2. Auto- somal (or typical Mendelian) inheritance, depending on genes in autosomes;
3. Y-chromosome inheritance, depending on genes in Y.

This dependence of characters on genes in the different chromosomes results, in later generations, in the production of certain numerical ratios among the individuals manifesting different characters. Of these the most common is the 3 to i ratio occurring in generation F2, in autosomal inheritance.

In the present chapter we proceed to a systematic examination of the diverse inheritance types and ratios, in their relation to the characteristics manifested.

The most general principle underlying this matter, we shall find, is the following:

The type and ratio of inheritance of any character is a relative matter: it depends upon the relative genetic constitutions of the two parents and on the characters manifested by them. More specifically, the type and ratio shown in inheritance by any character depends on what other character or characters it is mated with.

Many examples of these relations are found in the following pages.

In the early study of genetics it was commonly believed that each separably heritable characteristic, such as red eye or vermilion eye or rudimentary wings in the fruit-fly, depends on one gene only, and follows in its descent the distribution of that gene. Any characteristic that gave in heredity the simple Mendelian ratio (3 dominants to i recessive in Fa), or the simple sex-linked ratio, was considered a 'unit character', and was said to show 'unit character inheritance', depending supposedly on but a single 'factor' or gene. (The word factor came into use before the nature and location of the genes were known; it is still much used as a synonym for gene.) This single factor was supposed in some way to represent in the germinal material the character to which it gave rise. It was thus called 'the factor' or 'the gene' for that character, a form of expression which still persists, though its implications are erroneous.

Later it was discovered that this'conception is a mistake, since every characteristic is produced by the interaction of many genes, and can be altered by changing any one of them. The reason why certain characteristics give in heredity the simple Mendelian or sex-linked ratio is that the two parents differ in but one pair of the many gene pairs that influence the character. If the two parents differ in more than one of the gene pairs that influence the character, more complex ratios result. Furthermore, the type of inheritance shown by a given character (whether sex-linked, autosomal, or of the Y type) depends on the location of the gene pair or pairs in which the two parents differ. These important relations, with the general principles underlying them, are illustrated in the following sections, which take up in order the types of inheritance, and the ratios that result when the parents differ in one or more gene pairs, whether in X, the autosomes, or in Y.

1. Parents differing in a single pair of autosomal genes. As seen in the preceding chapter, the red programmes of the eye in the fruit-fly depends on at least 19 gene pairs scattered in various parts of the four pairs of chromosomes. An individual with the usual red eyes has all these genes in the normal or unmodified condition. Suppose that such an individual is mated with another that has all but one of these 19 gene pairs in the normal condition. This individual has the two genes of the pair at II, 54.5 so modified as to cause the eyes to be purple in programmes. We may call these two modified genes aa, while the corresponding two unmodified genes may be designated AA. The chromosomes of the two parents will then be represented as in Fig.they differ only in the genes at II, 54.5.

   When germ cells are formed in the usual way by such individuals (one chromosome of each pair passing to each germ cell), and the germ cells from the two parents are mated, we of course find that all the individuals (zygotes) of Fi have both the dominant normal gene A and the recessive gene a, so that they have the normal red eyes.

   When two of these individuals Aa of Fi form germ cells, and are mated in the usual way, they yield in Fs individuals of three different genetic constitutions, in the proportions AA + zAa + aa, or 3 dominant red to I recessive purple. The red and purple are inherited in the typical Mendelian or autosomal ratios for a single-gene difference, or as 'unit characters'.

   (It is obvious that to obtain the correct constitutions and ratios in later generations no account need be taken of the parts of the genetic system that are alike in the two parents. All that is required is to represent by different letters the genes that differ in the two parents. The two parents may thus be called AA and aa_9 the Fi generation Aa, the Fa generation AA-f zAa

+ aa. It is only when the conditions are complex that it is necessary to represent the chromosomes.)

2. Parents differing in a single pair of genes located in the X-chromosome.The same character red eye-programmes, just considered, gives in other cases a different type of single gene inheritance. This occurs when the recessive parent differs from the dominant redeyed parent in a single gene or gene pair that lies in the X-chromosome.

Let the red-eyed individual be a male, the recessive parent a female having eosin eyes. The female thus differs from the male in having the recessive eosin gene in her X-chromosomes, at the locus I, 1.5 Again represent the two recessive genes by aa. The corresponding normal gene in X will be called A, and as the male has but one X and therefore one gene A, it may be represented as AO (in which O represents the absence of this gene). We now have the situation:-

Males Females

Parents -> T> j ^ r? ýÿ

Red eyes Eosin eyes

Germ Cells A and O a and a

aO f Aa

Eosin eyes \ Red eyes

Thus the two eye colours now give sex-linked inheritance, with 'criss-cross' inheritance in Fi: the sons have eosin eyes, the daughters red eyes. If we mate together the males and females of Fi, forming the germ cells as usual, we obtain for the constitution of Fa in this case:

Sons AO -f aO; Daughters Aa -f aa

That is, in Fa, half the sons and half the daughters have red eyes; the other half, eosin eyes.

Thus the same character, red eye-programmes, which in the former case gave 'unit character' or 'single factor' inheritance of the autosomal type, gives in this second case 'unit character' or 'single factor' inheritance of the sex-linked type, with very different ratios. The way that the character red is inherited depends on what it is mated with.

Similar analyses can be given for all cases in which the two parents differ in but one pair, out of the many gene pairs on which depends the character examined. Thus in probably all cases in which characters give 'unit character', or 'single factor' inheritance ratios, the character is in fact affected by many diverse gene pairs, but only one pair differs in the two individuals that mate. The expressions 'unit character' and 'unit character inheritance' are still much used, but they should always be interpreted as meaning 'unit difference'; that is, as inheritance resulting from difference in a single pair of genes.

But if the parents differ in more than one pair of the genes on which the given character depends, the simple Mendelian or sex-linked 'unit character' ratios will not be produced, but more complex ratios. This may be illustrated for the same character, red eye-programmes, as follows:

3. Parents differing in two pairs of autosomal genes, in different chromosomes.The red-eyed parent is mated with an individual that has two of the 19 eye-programmes genes modified; one pair in chromosome II at 54.5 giving recessive purple eyes; another pair in chromosome III at 48.0 giving recessive pink eyes.

As before, we need to take account only of the genes in which the two parents differ. Call the two recessive genes that give purple eyes aa, the two that yield pink eyes, bb. Then the corresponding dominant genes in the other parent may be called AA and BB, these giving red eyes. It is necessary to keep in mind the eye programmes in the different types of individuals. Those having only the two recessive genes aa have purple eyes; those with only bb have pink eyes; those with both aa and bb have purple-pink eyes. Those with neither pair of recessive genes have red eyes; these are AABB.

When the red-eyed parent AABB is mated with the purple- pink-eyed parent aabb, for parents differing in two autosomes. The Fi generation all h^ve the constitution AaBb, and since A and B are dominant, all the individuals of Fi have red eyes.

Now mate together individuals of Fi; they yield the combinations. And as there shown, if we classify these as to the characteristics that are manifested (that is, as to what dominant genes they contain), the proportions are the following:

gAB + 3Ab + 3aB + lab

All those that contain both the dominant genes A and B have red eyes. Those that contain the dominant A but not the dominant B have pink eyes. Those that contain the dominant B but not the dominant A have purple eyes. Thus the individuals with different eye colours occur in F2 in the proportions: nine red-eyed (AB) to three pink-eyed (Ab) to three purple-eyed (aB) to one pink-purple-eyed (ab). Or we can classify them as 9 red-eyed to 7 not red.

Thus in this case the character red gives in F2 still different proportions from those that were found in the two previous cases. It now gives 'two-factor' autosomal inheritance. It illustrates anew the fact that the type and ratios of inheritance given by any character depend upon what it is mated with. 4. Parents differing in two gene pairs in the same autosome.ýÿWhen the gene pairs in which the parents differ are located in the same chromosome, the ratios are altered by the fact that genes in the same chromosome usually go together to the descendants; that is, they show linkage. An ideal case will illustrate the results.

Let one parent carry the two recessive gene pairs aa and bb, located in the same autosome. The corresponding dominant genes, in the other parent, may be called as usual AA and BB. Their Fi offspring will then be a heterozygote of the type shown in Fig. 45, Fi. Its two chromosomes may be designated A ýÿ B and a ýÿ b.

Assume that the distance between the two gene pairs is such that exchange takes place in one-tenth of all cases, the exchanges yielding chromosomes carrying A ýÿ b and a ýÿ B, in place of the original A ýÿ B and a ýÿ b. Then in nine- tenths of all cases the chromosomes retain the original constitution, A ýÿ B and a ýÿ b] in one-tenth they have the changed constitution A ýÿ b and a ýÿ B.

Now, when these chromosomes separate into different gametes, there will be formed four types of gametes having respectively the chromosomes A ýÿ B, a ýÿ b, A ýÿ b, and a ýÿ B and these four types will occur in the proportions just given. That is, of the entire number of gametes, the four types occur in the following proportions:

ýÿ45AB + -O5Ab + -05*6 + *45ab

Since it is only the relative proportions that are important, all terms of this expression may be multiplied by 20 to make them integral. They then yield the proportions

gAB + i Ab + laB -{-gab

These different types of gametes mate in the above proportions with the gametes from the other parent. In such an organism as Drosophila, in which exchange of genes occurs only in the female, it is only the ova that will exist in the above proportions. The male, in which no exchange occurs, will produce but two sorts of gametes in equal proportions, those containing the chromosome A ýÿ B, and those containing the chromosome a ýÿ b. The sperms therefore are in the proportions:

Or, what is the same, The zygotes of Fa will then be produced in proportions given by mating each of the two types of sperms AB and ab with each of the four types of ova in the proportions in which they occur. The results are given simply by multiplying together algebraically the two expressions:

(9AB +1 Ab + i aB + gab) (AB + ab)

This yields for the constitution of the generation Fa the following:

gABAB + lABAb + lABaB + iSABab + lAbab + laBab+gabab

Since the capital letters represent dominant characters, the characters A, B, a and b will be manifested in Fa in the following proportions:

agAB +1 Ab + laB + gab

In case crossing-over (or exchange) occurs equally in both sexes, the constitution of the Fa generation will be given by the product of the two expressions:

(gAB + iAb+ laB + gab) (g AB + i Ab + i aB + gab)

This yields for the constitution of the Fa generation individuals of different genotypes in the following proportions:

81ABAB +1 SABAb +18 ABaB + 16a ABab + i Ab Ab + a AbaB + 18 Abab + i aBaB + i SaBab + 81 abab

If we classify these as to whether they manifest the characters represented by A, B, a or b, we find the proportions to be as follows:

a8iAB+igAb+igaB + 8iab

The method of computing the results may be generalized as follows:

Call the two pairs of dominant genes AA and BB, the two pairs of recessive genes aa and bb. Then the one parent has the two chromosomes AýÿB and AýÿB, the other parent aýÿb and aýÿb. The individuals of Fi then all have the two chromosomes AýÿB and aýÿb,

ao6 Let r=the exchange (or cross-over) ratio (the proportion of cases in which the chromosomes exchange genes, in the germ cells produced by Fi).

Then i -r=the proportion of cases in which there is no exchange.

Then in the individuals in which exchange occurs (one or both sexes) gametes are produced in the proportions:

(i -r)AB + rAb + raB+(i -r)ab (i)

If in one sex there is no exchange, the gametes from the individuals of that sex are of but two kinds, in the proportions:

AB + ab (2)

(Since in these expressions (i) and (2) it is only the relative proportions that are important, either or both may be multiplied through by any convenient factor to make the numbers integral; the factor need not be the same for both.)

The proportions of the different types of individuals (zy- gotes) in F2 are obtained by multiplying together the expressions for the proportions of the gametes in the two sexes. That is, the proportions in F2 are, if crossing-over occurs in one sex only:

[(i -r)AB + rAb + raB+(i -r)ab] [AB + ab] (3)

While, if exchange occurs in both sexes, the proportions in F2 are:

[(i-r)AB + rAb + raB+(i-r)ab][a]

The value of r in case of different pairs of genes is determined by extensive observation of the results of experimental breeding. For different pairs of genes of Drosophila, many such values of r are given in the various publications of Morgan, Bridges and Sturtevant.

4. Parents differing in two gene pairs of the X-chromosome.ýÿIn this case, ratios diverse from those given by auto- somal genes will be produced, owing to the different method of descent of X. The two genes will show sex-linked inheritance, with linkage. It will be found an interesting exercise to give some definite value to r, the exchange or cross-over ratio, then to work out formulae for computing the proportions of the different types of zygotes in Fa.
5. Parents differing in two pairs of genes, one pair in X, the other in an autosome.ýÿIn this case complex ratios result for the composition of Fa. These can be worked out by representing in diagram the two kinds of chromosomes in the two parents and putting them symbolically through the processes of gamete formation and fertilization.
6. Parents differing in several pairs of genes located in the same or different chromosomes (autosome or X).ýÿIn such a case the computation of inheritance ratios is very complex. It must be done by representing in diagrams the different chromosomes and the position of the genes within them, then putting them symbolically through gamete formation and fertilization, taking into consideration the known exchange ratios.
7. Genes in the Y-chromosome.ýÿIn most organisms, seemingly few genes,· or none, occur in the Y-chromosome. As they do occur in rare cases, it will be well to indicate the resulting ratios for certain different cases.

   (a) A dominant gene A in Y, corresponding to a recessive gene a in the X-chromosome. This case is realized in Droso- phila, in cases in which the X-chromosomes contain the recessive gene bobbed, which shortens the body bristles, while the Y-chromosome contains the dominant normal or wild- type gene causing the bristles to be of normal length. The two parents are then:

Females Males

Parentsýÿ aa aA

Bobbed bristles Normal bristles

Here the a's are in the X-chromosomes, the A in the Y- chromosome. Since Y goes only to males, the Fi generation will be:

Females Males

aa aA

Bobbed bristles Normal bristles

This situation will continue in later generations: all the males dominant, all the females recessive.

b. The dominant gene may occur in some of the X's, as well as in Y. Suppose that the X of the male has the dominant gene A, as does also the Y, while the X's of the female have the recessive gene a. Representing the X that contains the dominant gene by A, while the Y that contains it is represented by Ay the following is the situation:

Females Males

Parents: aa AA

Recessive Dominant

Fi aA + aA aA + aA

Dominant Dominant

Fs Aa + aa AA + aA Dominant Recessive Dominant

In such a case, therefore, all the males are dominant in every generation. But in F2, half the females are recessive, and recessive females will appear also in later generations.

c. If Y contains a recessive gene, while the X's contain the corresponding dominant, the effects of the recessive gene will never be manifested, since both sexes carry the dominant X. Thus Y might contain many recessive genes, none of which would ever be manifested.

d. In certain fish, crossing-over of genes from X to Y or vice versa has been described as occurring in rare cases. This of course changes the method of inheritance. Details need not be entered into here, since such cases are very infrequent.

9. In some cases, as will be shown in later chapters, when the chromosomes are broken into pieces by radiation, a piece of an X-chromosome may become attached to an autosome, remaining thus attached in later generations. In such a case the genes of this piece of X, which normally give sex-linked inheritance, after the breakage give autosomal inheritance. Similarly, a piece of an autosome may become attached to X, its genes thereafter yielding sex-linked inheritance.

Certain important special conditions, some of them classifiable, as to certain features, under cases already considered, are worthy of special consideration. These are the following:

10. Both parents recessive for a single pair of genes.ýÿWhen two such recessive parents are mated, the nature of the characters shown by the offspring differs in different cases, depending on whether the recessive genes of the two parents are or are not alleles; that is, on whether they are or are not at the same locus of the same chromosome. A. The recessive genes in the two parents are alleles, being located at the same loci of the same chromosome.

In this case no corresponding dominant gene is present. All the offspring and descendants are therefore recessive. This is the result when the two parents are alike in their recessive characters, as when two eosin-eyed parents are mated. It is also the result when the characters of the two parents differ, though due to different alleles of the same gene; for example, when parents with eosin eyes are mated to parents with buff eyesýÿboth of these characters being due to modification of the gene at I, 1.5 in Drosophila. In such a case the heterozygotes, containing the two different modifications of the gene, are either intermediate in character, or approach one of the recessive parents more closely than the other.

The recessive genes in the two parents are not alleles, but are modifications of genes at different loci. In such cases the offspring (Fi) show, not the recessive characters, but the normal or dominant characters. The reason for this is, A, that each parent has the dominant allele for the recessive character of the other parent.

If the two pairs of recessive characters of one parent are bb and those of the other parent are ee, then the parent carrying.This is an extremely important relation. It holds both when the two recessive characters affect the same structure or function, and when they do not. Thus, if flies with eosin eyes are mated with those having vermilion eyes, the offspring all have normal red eyes. Or again, if vermilion eyes are mated with purple eyes, the offspring have normal red eyes. In such a case the recessive characters of the two parents may be indistinguishable; but, if they are not alleles, the offspring have the dominant character. Thus, the eye colours garnet (gene at I, 44.4) and purple (at II, 54.5) are alike, but if garnet and purple parents are mated, the offspring have normal red eyes. Similarly, if rudimentary wings are mated with vestigial wings, the offspring have normal wings.

If the recessive characters of the two parents are quite diverse (and not alleles), the same rule holds: purple-eyed flies mated with vestigial wings yield offspring

(Fi) that are normal both for eyes and wings.

11. Duplicate genes.ýÿIn certain cases a dominant character is produced in case a dominant gene of either of two pairs, AA or BB, is present, while the corresponding recessive character occurs only when both these dominant genes are lacking, so that in their place are two recessive pairs aa and bb. Such cases give special inheritance ratios; the proportion of dominants to recessives is 15 to i instead of 3 to i. These are known as cases of duplicate genes.

We have before mentioned a case of this sort, described by G. H. Shull in the common weed Bursa bursa-pastoris, or shepherd's purse. Some races have dominant triangular pods, while others have recessive ovate pods. It appears that the race with ovate pods carries two recessive gene pairs, aa and bb, present in different autosomes. The race with triangular pods has two corresponding dominant pairs of genes, AA and BB, and the presence of a dominant gene of either of these pairs suffices to cause the production of the triangular pods. When these two races are crossed, therefore, the following is the situation:

Parents: AABB aabb

Triangular Ovate Fi: AaBb Triangular

When these Fi individuals are bred together they of course produce in Fa the sixteen sets of individuals of the constitutions shown. As examination of this table shows, 15 out of the 16 contain either A or B or both; hence these 15 have triangular pods. Only one of the 16, that with the constitution aabb, contains neither A nor B; this one has therefore ovate pods. Thus in such cases the ratio in Fa is 15 dominant to i recessive.

A considerable number of such cases are known, yielding the ratio 15 to i. They occur mainly in plants. In certain cases there are indications of a ratio of 63 dominants to i recessive in Fa. This is evidence that the dominant condition is produced if any one of three dominant genes. A, B or C, is present, while the recessive condition occurs only in individuals of the constitution aabbcc.

12. 'Blending inheritance.' Intermediate characters. 'Multiple factors.'

There are many cases in which a cross between two differing stocks results in progeny (Fi) that are intermediate between the two parent stocks, and this intermediate condition may continue, in large measure, in later generations. Such intermediate or blending inheritance is found mainly in characters that grade gradually from one extreme to another: dimensions, degrees of programmes, and numerical characters, such as numbers of leaves in certain plants. A cross between large and small individuals commonly results in progeny that are intermediate. Crosses between whites and negroes give offspring that are intermediate in programmes.

It was at first supposed that such 'blending' inheritance is of a fundamentally different type from Mendelian inheritance, that is, from inheritance depending on differences among the genes. This was an almost necessary conclusion when it was believed that each Mendelian character depends exclusively on a single factor or gene. But since it has been discovered that every character depends on many genes, and that parents may differ in one or many genes affecting a particular character, so-called blending inheritance has lost its unique character, and has been recognized as resulting, like other inheritance, from differences between the genes of the two parents. Many intermediate conditions have been discovered between ordinary single-gene inheritance and the phenomena of blending inheritance. These, with the evidence for the dependence of blending inheritance on gene differences, are set forth in the following.

A. Lack of dominance: heterozygotes intermediate.ýÿIn some cases, as we have seen on earlier pages, when two parents differing in a single gene pair are mated, neither character shows dominance. The offspring (Fi) from the mating are intermediate between the two parents, or show some blend of the two differing characteristics. That is, if the genes of the two parents are AA and aa, the heterozygotes Aa are not like either parent, but show a blend of their characteristics.

This is the usual case, as before seen when parents showing two diverse recessive alleles are mated in Drosophila. Thus if eosin eye is mated with buff eye, the progeny (Fi) show eyes of intermediate programmes; if vestigial wings are crossed with an tiered, the progeny (Fi) show wings of intermediate type.

Many cases of this sort are known in various organisms. One of the best known is that of the Andalusian fowl. Here a black individual (AA) mated to a white one (aa) gives in Fi individuals (Aa) that show a somewhat intermediate programmes known as blue. Similar cases are known for plant and animal colours of various sorts. The heterozygote in such cases is frequently not closely intermediate between the parents; in different cases it shows various intergradations of the parental characters.

In such cases of lack of dominance, the ratios in which the different characters appear in later generations are changed; the familiar 3 to i ratio does not appear. If the parents are designated AA and aa, then the heterozygotes Aa of Fi are all intermediate. When these are bred together, they yield the usual proportions:

AA + 2 Aa + aa

Thus there are in F2 three grades, one (AA) like one parent, one (Aa) intermediate, and the third (aa) like the other parent. And these three grades occur in the proportions i: 2: i. Here we have a simple example of what in more complex cases is known as blending inheritance.

B. Dependence on differences in several or many gene pairs, having quantitative effects: 'Blending inheritance', 'Multiple factors'.

As before seen, any characteristic is affected by many different genes, though if the parents differ in but a single one of these gene pairs, the result is 'single-gene' or 'unit-character' inheritance. But the parents may differ in two or more of the genes affecting the characteristic, then yielding inheritance in various complex ratios. The gene pairs in which the parents differ may have qualitatively diverse effects, as in the case of eye programmes in Drosophila, or coat programmes in rodents.

But in other cases the different genes affecting the character differ only quantitatively in their effects. Thus, in peas, one of the cases studied by Mendel was a cross between tall and dwarf peas. These differed in a single gene pair; one of these (AA) gave tall plants; the other (aa) gave short plants. The gene A for tallness was dominant, so that the heterozygotes Aa were tall.

When such cases of quantitative difference in effects are combined with the production of intermediate conditions in the heterozygotes, and particularly when the parents differ in several gene pairs, or 'multiple factors', the result is to produce many gradations of the characteristic, so giving rise to what has been called blending inheritance.

Such inheritance, as before remarked, is seen particularly in such matters as dimensions, numerical characters, gradations of programmes, and the like. But, as just seen, in some cases such characters depend on single-gene differences between the parents; they then give typical Mendelian or sex-linked ratios in the descendants.

Dimensions thus giving typical single-gene inheritance are: tallness and dwarfness in peas, in sweet peas, in Antirrhinum (the snapdragon), in tomato plants. Long and short styles in Oenothera; long and short wings in Drosophila; long and short hair in various rodents; long and short legs in Dexter- Kerry cattle; long and short fingers in man. Differences in programmes that give typical Mendelian inheritance are numerous in Drosophila.

In other cases of difference in dimensions or programmes, or the like, the progeny are intermediate between the parents, and there may be many different gradations. This is the case for stature in man, for ear length in rabbits, for depth of programmes in negro-white crosses, for many dimensions in organisms.

The key to the understanding of these cases was given by the discovery that, when such crosses occur, the descendants in the ¥2 generation are much more varied than those in the Fi generation. This is, of course, necessarily the case if the characters depend on differences in genes. When parents differ in a single pair of genes AA and aa, the offspring all have in Fi the same constitution, Aa. But in Fa there are three different constitutions AA + 2Aa -f aa, so that Fa is much more varied than Fi. Again, suppose that the parents differ in two pairs of genes, so that one parent is AABB, while the other is aabb. Then the Fi generation all have the constitution AaBb, and are therefore all alike. But in the Fa generation there are nine diverse combinations of the genes, as shown they therefore show much more diversity than occurs in Fi.

An illustration will show how this works. Suppose that two races differ in size, in consequence of a difference in two pairs of genes. The larger race carries the genes ABAB, while the smaller one has the corresponding genes abab. Assume that each one of the four genes ABAB increases the size by a certain amount above that of the race abab. When the two races are crossed, the Fi generation all have the constitution ABab. Having two of the four genes that increase the size, they are intermediate in size between the two original races, and, as all have the same constitution, there will be little variation in size.

Now mate together two of these individuals AaBb. These will give for the Fa generation the combinations shown. Of these, those that have four of the genes ABAB will be largest, those with but three of them will be smaller, and so on down to those having none of themýÿthe individuals with constitution abab. There will thus be a series of graded sizes.

To determine the relative number of individuals in the different grades, from smallest to largest, the constitutions having respectively o, 1,2, 3, or 4 of the 'size genes' ABAB. The result is as follows:

Number of size genes 012 34

abab abaB abAB ABAb ABAB

i aBab aBAb AbAB i

abAb ABaB ABaB

Abab ABab aBAB AbAb 4 aBaB

Thus there are in F2 five grades as to size. The most numerous individuals (6 out of 16) have the intermediate size, with but two of the 'size genes. There are 4 a little smaller and 4 a little larger, with a single individual in each of the extreme classes. The numbers of individuals in the different classes are given by the binomial series resulting from expanding the expression (i + i).

Similarly, if the two original races differ in three pairs of genes in different chromosomes, so that one is AABBCC, while the other is aabbcc, the Fi generation will have the uniform constitution AaBbCc, and so will be intermediate between two races. The F2 generation will be found to have seven size gradations containing respectively o, 1,2, 3, 4, 5, or 6 of the 'size genes' AABBCC, the different grades showing the following relative numbers of individuals:

No. of size factors 0123456

No. of individuals i 6 15 20 15 6 1=64

Here the relative numbers of individuals in each of the grades is given by the expansion of the binomial (i + i). The relative numbers for any other number of 'size genes' can be written out directly. For four pairs the series is given by (i + i), and for n pairs it is (i -f- i).

Thus when there is a cross between races differing quantitatively in size, depth of programmes, or the like, the difference being due to a diversity in several pairs of genes, the general results are as follows:

1. The Fi generation is uniform (except in so far as there are variations due to environmental conditions), and is intermediate between the two parental types.
2. The F2 generation is varied in constitution. It consists of a series of grades, varying from the larger to the smaller parent. If the number of gene pairs in which the original parents differ is n, the number of grades is zn 4-1.
3. The number of individuals is largest in the intermediate grades, and decreases toward the more extreme grades, in both directions.
4. The relative numbers of individuals in the different grades, from smallest to largest, are given by the expansion of the binomial $(i\ \text{-f}\ i)^n$.
5. Since most of the individuals fall in the intermediate grades, the general effect is to give the impression that F2, like Fi, is intermediate, so that the inheritance appears to be of the blending type. This is particularly true if only small numbers of individuals are available for study; most or all of these fall in the intermediate classes.
6. But careful study, when large numbers of individuals are available, shows that the F2 generation is much more variable than the Fi generation. It reveals also the presence of a few extreme individuals, differing little or not at all from the two parental types.

This greater variability in F2 as compared with Fi, in cases in which the parents differ quantitatively, has been found to occur in great numbers of cases, of which the following may be mentioned: size and shape in gourds, squashes, beans; length of the ears in maize programmes, size of seeds and of petals in flax time of flowering in peas and in cotton size in ducks winter hardiness in cereals skin programmes in negro-white crosses size in rabbits and in many other cases. There appears to be no known case of 'blending inheritance' in which the Fa generation is not more variable than the Fi generation, so that it appears clear that in all such cases the inheritance is based on differences in a number of genes ('multiple factors'), each of which has a quantitative effect on the characteristic.

7. The later generations, produced from the interbreeding of the F2 individuals and their descendants, show much the same relations as are found in F2. Most of them are intermediate, though with a number of different grades. A few are extreme; these, however, are so rare that they are observed only when large numbers are bred.
8. The intermediate individuals in F2 or later generations, when interbred or self-fertilized, usually do not breed true; they produce a number of different grades, so that their offspring are variable. This is because the intermediate individuals:
   a. Include a number of different grades, with different constitutions;
   b. Many of them are heterozygotic. The commonest intermediates are those with the constitution AaBbCc, etc., others show such constitutions as AaBBCc, AabbCc, and so on. These when interbred must give individuals of many different constitutions.
9. The rare extreme individualsýÿfor example, those that are very large or very smallýÿwhen bred separately commonly 'breed true', producing offspring like themselves. This is because such extreme individuals are:

a. All of the same constitution,

b. Homozygotic. The extreme individuals at one end of the scale have the constitution AABBCC, etc.; when two such homozygotes are bred together, they can produce only homozygotes like themselves. The extreme individuals at the other end of the scale have the constitution aabbcc, etc.; these also are homozygotes and can therefore produce only offspring having the same constitution as themselves.

10 Results of Selection.ýÿThe peculiar relations set forth in paragraphs (8) and (9) above come out strongly in experimental work on the results of selection, in relation to quantitative differencesýÿsizes, weights, yield of crops, and the like. Attempts were early made to improve the breeds of cultivated plants and animals by selecting and breeding together those of the better grades. The results were found to be as follows:

a. If parents above the average are selected, these give offspring above the average, but varying.
b. If from these offspring those of higher grade are selected, again these give offspring of higher grade, but varying. Thus by repeated selection of the higher grade individuals in successive generations, progressive improvement is produced, 'selection is effective. Such selection in the opposite direction is equally effective.
c. But after this has continued for a number of generations, improvement ceases; no further progress is made. Selection ceases to be effective. The individuals of the highest grade, when bred by themselves, produce offspring like themselves, without genetic variation, and the same is true for the individuals of lowest grade.

This cessation of the effectiveness of selection after a certain grade was reached was unexpected, and gave origin to much discussion. But after it became clear that in such cases multiple factors are at work, in other words that the various grades depend on the number of certain types of genes that are present, the matter was cleared up. Selection is effective so long as it deals with the intermediate grades, which are of various different genetic constitutions, and are frequently heterozygotic. But as soon as the homozygotic individuals AABBCG, etc., or aabbcc, etc., are reached, no further progress can be made, since these can produce only offspring having the same constitution as themselves.

On the whole it is fully demonstrated that so-called blending inheritance is inheritance resulting from gene differences, such as yield ordinary Mendelian inheritance; but in blending inheritance the parents differ in several or many pairs of genes, each gene having a quantitative effect.

Chapter 11

Relation of Characteristics to Environment

ITS INTERACTION WITH HEREDITY

We have seen in earlier chapters that all characteristics of organisms, of whatever kind, depend on genes and can be altered by altering genes. This is no more than to say that all properties of organisms, as of inorganic bodies, depend on the materials of which they are made, and can be altered by altering those materials. Changing the genes alters the materials of which organisms are made, and thus may alter any of their characteristics.

But this leaves open the possibility, that other things also may alter any or all of the characteristics. The fact that alteration of genes changes characteristics does not preclude their alteration by the conditions to which the organism is subjected during its life and development. In the present chapter we examine the action of environmental conditions on characteristics and their interaction with the genetic constitution.

To form a correct conception of the interaction of environment and genetic constitution, one must look at the method of manufacture of an organism: that is, at the nature of the process of development from an egg to an adult. To certain features of this we now turn.

Relation of Gene Action to Internal Environment in Individual Development

The organism starts as a single cell, the fertilized egg cell. This contains the great number of genes, a thousand or so, groupec} together in the visible chromosomes.

These chromosomes are seen to go at once to work. They are embedded in a mass of material, the cytoplasm, forming the body of the cell. They take up material from the cytoplasm, so that they swell, enlarge, become vesicles, become crowded together to form what we call the nucleus. They chemically change this cytoplasm; they give it off again into the cell, so that the chromosomes are again minute condensed bodies. The changed cytoplasm that they have given off into the cells alters the nature and structure of the cell, often in a strongly marked, visible way.

This process of changing the cytoplasm by the action of the genes is the fundamental thing in development. The genes repeat this process over and over againýÿtaking in cytoplasm, modifying it, giving it off in changed conditionýÿleaving the genes themselves unaltered.

While this is going on, the cell divides into 2,4,8 cells, and so on. The different cells receive diverse kinds of the cytoplasm that has been worked over by the genes. So the different cells gradually become diverse. As the cells divide, each chromosome divides, each gene divides, and half of every gene goes into each cell, where it grows again to the original size. Thus every cell of the body contains all the genes, the complete set. The cells differ in their cytoplasm; they do not differ, as a rule, in their genes.

The cells keep getting diverse, in this way, till some produce bones, some muscles, some nerves, some eyes, some hands, till finally we have the complex individual with all his parts and functions.

How does it happen that the different parts of the body become diverse? Since all the genes are present in all the cells, why do not all the cells change in the same way, instead of becoming different?

Not very much is known about this; it is one of the darkest questions of biology. But some things are known about it that are of great interest for the question as to the effects of environment. Look at one or two experiments.

The original single cell divides into two cells A and B.

These two cells later produce the two halves of the body; that is, one produces the right half, the other the left half.

Why does the cell A produce only the left half of the body instead of the whole body? Try separating the two cells. This can be done, with some difficulty, in certain organisms. Now we find that this cell A, when separated, produces, not the left half of the body, but the entire body, an entire individual; and the right-hand cell B will do the same.

So it is clear that what each cell produces depends on its environment, on its relation to the other cells. Separate the two; then the genes in each produce an entire individual, in place of half an individual.

This sort of thing turns out to be typical for development. Every cell contains as a rule all the genes, and so far as genes are concerned could produce an entire individual. But after a while, through the continued activity of the genes, many different substances have been manufactured from the cytoplasm and are located in the different cells. Things have now changed so much, and have become so fixed, that the genes can no longer start anew from the beginning. At such a later stage therefore a single separated cell will no longer produce an entire organism, or any required part of the organism, although it still contains all the genes. With age the cells become fixed in their ways, as old individuals do.

But this condition in which any cell can produce any one of many different parts of the body, depending on circumstances, may continue into a rather late period of development, particularly in the vertebrates. At a certain time the egg of such a creature as the frog has become a mass of small cells. Under normal conditions, when we examine this mass, we can predict what part of the adult each part will produce. Cells here will produce the brain, there at the sides the eyes, here the ear, there the spinal cord, here parts of the skin. If we leave the egg to itself, these are indeed the parts that will be produced.

But this is not because each cell can produce only that part and nothing else. For it is possible to cut off a disk of cells from the egg and turn it around, so that what was in front is now behind, what was right is now left. And now we find that all the cells change their method of action. The cells that would have produced brain now produce skin; what would have yielded skin now yields brain or eyes; what would have given spinal cord now produces cerebral hemispheres. What happens is that from a certain spot on the eggýÿa recognizable spotýÿan organizing influence starts out, so that this spot is known as the organizer, or the organization centre. This organizing influence, whatever its nature, creeps from cell to cell, causing each cell to alter internallyýÿthrough the interaction of its genes and cytoplasmýÿin such a way as to produce the structures of the embryo. Each cell that is reached later transforms in such a way as to fit the cells that have gone beforeýÿin such a way as to make the next proper part in the pattern of the body. At a certain point the cells transform into spinal cord, the next ones into medulla, those next into midbrain, the next into forebrain, those at the sides into eyes, farther forward into skin. This organizing influence passes through, in the same way, whichever cells are present, so that it is clear that any of the cells can produce any of the parts that are to be produced. If one turns them side by side or end to end, the results are the same. Any cell can produce almost any of the parts of the body.

That is, then, from the beginning the cells adjust themselves to their surroundings, to their environment. What part of the body each cell becomes is determined, not by the genes it containsýÿfor each contains all the genesýÿbut by the conditions surrounding it. The genes within the cell are enormously sensitive to their environment, to the cells surrounding them. They alter their method of action to fit the situation, producing whatever is the next thing in the body pattern. Development is fundamentally adjustment to environmentÿÿin this case, to cellular environment.

Of course, what the genes may produce is not without limits. The genes of a frog produce the parts of a frog's body; the genes of a man produce the parts of a man's body. The genes of a Chinese produce the body of a Chinese; the genes of a Caucasian, the body of a Caucasian. Some genes produce an individual of light complexion; others, individuals of dark complexion. Just what the gene shall produce in development is always dependent both on the nature of the genes that are at work and on the conditions in which they find themselves. And in this they give a true picture of the interactions of heredity and environment.

The dependence on the conditions surrounding the cells, of what the genes within the cells produce, is shown in certain other ways, in the production of differences of sex. As there seen, in many organisms the two sexes begin life with different chromosome combinations: one sex has two X-chromosomes, the other but one. It is to this original difference in gene combination that the later sex difference is ultimately to be referred. But the original difference in chromosomes acts through intermediate chemical products, and it is these diverse chemical products that cause the bodies of the two sexes to develop differently. In the embryo containing but one X-chromo- some a certain kind of hormone is produced, in that having two X-chromosomes another sort of hormone. These two diverse hormones then circulate through the two bodies; they act diversely in the two, causing one to produce the characteristics of the male sex, the other the characteristics of the female sex. Body cells of either type may produce either type of characteristics, depending on the hormone that acts upon them. That is, what their genes produce depends on the chemical conditions that surround them. Diversity of sex, like diversity in tissue differentiation, is produced through adjustment of the cells to their environment; in this case to hormonal environment.

It has been discovered that many other diversities between individuals, and between parts of the body, are produced by other kinds of hormones, by hormones secreted by the thyroid gland, by those from the hypophysis, and from other parts. Details cannot be entered upon here, but what they agree in showing is this: The genes first act to produce different chemicals. Then these diverse chemicals cause the cells to develop in certain particular ways, so as to give rise to diverse parts. By this continued interaction of genes with cytoplasm, of cells with each other, of cells with their surroundings, there is finally produced the entire body with all its parts and functions.

Thus in the process of development, adjustment of the cells to their environment, response to the surrounding conditions, play a fundamental part. The genes show themselves most sensitive to the conditions surrounding them.

Since genes act by producing hormones, and these hormones then determine many features of the later development, the question arises as to whether such hormones might not be provided in some other way. Could they be introduced into the body with the food, or by injection into the blood? If so, the characteristics produced would be controlled from outside. Or can the hormones be modified, or removed, in such a way as to change the characteristics of the individual?

As before seen this can indeed be done. The hormone produced by the male embryos (those having the single X-chromosome) may be transferred to the embryo that would otherwise produce a female, and cause it to produce male characteristics. In a similar way, the action of an imperfect thyroid hormone can be supplemented from outside. Individuals in which the genes supply a poor thyroid hormone fail to develop normally; in man they form the cretin, whose mental development is stunted. But this lack may be supplied by introducing the thyroid hormone with the nutrition; thereupon the development returns to its normal course.

In these cases products of the genes are transferred from one individual to another, where the same effect is produced as would result if the substance came from the operation of the individual's own genes. But a further step is possible. In some cases the chemicals controlling development are produced artificially and introduced into the developing body, where they produce the same effect as those produced by the organism itself. The active principle of the thyroid has been synthesized, and this synthetic product may be used in place of that produced by the genes, with the same effect on the development of the individual. Progress has been made in producing artificially other products of the genes, such as epinephrin, the secretion of the suprarenal body, and insulin, the internal secretion of certain cells of the pancreas. The production and use of hormones or endocrine secretions, and the control of development or function through their use, has become a vast chapter of physiology, which is enlarging rapidly. It all illustrates the fact that control of development and of characteristics is in large measure possible, and even at the present time is in a considerable degree practicable.

Altogether, therefore, it is clear that the processes of development and the determination of the nature of characteristics are not shut off from outside influence; but that, on the contrary, adjustment to the conditions is one of the fundamental features of development.

Thus far we considered mainly the role of environmental conditions in determining the different characteristics of different parts of the body. Do environmental conditions likewise play a role in determining the different characteristics of different individuals, so that individuals that have developed under different conditions have different characteristics? From what has just been set forth on the nature of development, this might be expected. We have seen that some of the characteristics developed depend on the hormones that are produced within the body. It is known that the production of some hormones is under the influence of the nervous system, and, through this, under the influence of outer conditions. This is the case in man with the hormone from the suprarenal capsules. This hormone affects behaviour, but little is known as to its effect on development. Other hormones, however, are known to affect development profoundly. Hormones from the hypophysis have a striking effect on growth. Differences in the quality and quantity of these hormones in different individuals result in the production of giants in some cases, of dwarfs in others, of normal growth in still others.

In another vertebrate a striking effect on development, altering all the adult characteristics, is known to be produced through the action of a hormone. The axolotl is a large salamander that has large red external gills, a tail flattened sidewise, for swimming, and other features that fit it for living in the water. It lives in the water thus all its life, becomes mature, produces eggs and young; there it dies; it is never anything but an axolotl.

But if it is fed on thyroid, it undergoes a transformation comparable to that which changes a tadpole into a frog. It

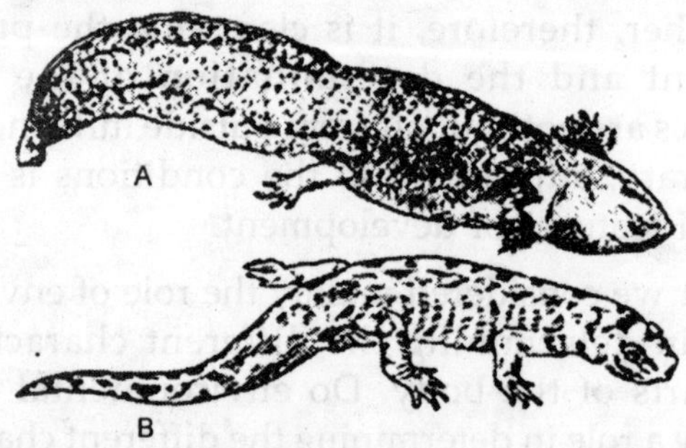

Fig.Axolotl (A), and Amblystoma (B), to show the difference inform and structure. Figs modified from those in Brehm's Thierleben.

loses its gills; its body form becomes greatly changed, its tail is no longer flattened sidewise. It becomes fitted for living on land, crawls out on the land, and is the creature known as Amblystoma. It now lives all the rest of its life under this guise, as a land animal; it becomes mature in this condition, producing eggs and young.

Of greatest interest from our present point of view is the fact that external conditions may induce this same transformation. If the axolotl is driven to cora £ out on l&rid under certain conditions of temperature and the like, it transforms into an Amblystoma, just as it does when fed on thyroid. There is little doubt that what these conditions do is to cause the thyroid of the animal to discharge its secretion into the blood, and that this induces the transformation. Thus the axolotl may have either one or two very diverse sets of characteristics, depending on the conditions that it meets during development.

CHARACTERISTICS AFFECTED BOTH BY GENES AND BY ENVIRONMENT

Any of the characteristics of the organism may be altered by changing its genes: this we have seen in earlier chapters. Many of the same features that can be altered by changing the genes can likewise be altered by appropriate changes in the environment. Characteristics do not fall into two classes, one exclusively hereditary (or dependent on genes), the other exclusively environmental; but any characteristic is affected both by the materials of which the organism is composed, and by the action of the conditions on these materials.

Yet, as we shall see, in practice some characteristics are more readily altered by environmental conditions than are others. Certain characteristics owe most of their peculiarities to diversities among genes. Others are readily affected both by genes and by environment. Still others depend mainly on environmental conditions.

Particularly illuminating are the cases in which characteristics depend in marked degree both on genes and on environmental conditions, so that they are altered by changes in either. A number of well-known cases of this kind will be presented, selecting those that are from a technical point of view fully known.

In that animal whose genetics is best known, the fruit-fly Drosophila, many of the characteristics that have been studied are defects or abnormalities which are sharply dependent on the presence of particular genes. Such a one is abnormal abdomen. This appears in the fact that the abdomen is ill formed, the segments not being regular nor sharply marked off. This abnormality is found to be due to a defect in a certain gene of the X-chromosome, so that it shows sex-linked inheritance, the abnormality being dominant. Unless that particular defective gene is present the abdomen is normal. But it is dependent too on the environment. Individuals with the abnormal gene, if developed in a moist atmosphere, have the abnormality. But if those same individuals are developed in a dry atmosphere, they have normal abdomens. To produce the abnormal abdomen, we must have first the particular kind of defective gene that yields it. But second, even if that is present, we must have, too, the moist atmosphere for development; without it the abnormality is not produced, whatever the gene.

Another character in the fruit-fly that acts in a similar way is what is called reduplicated legs. If a certain defective gene is present in the X-chromosome, the animals show a tendency to produce double legs, or even triple or multiple legs, in place of single ones. But this does not happen unless the young develop at a low temperature. To produce reduplicated legs, then, there must be present a certain type of gene, and also a certain external condition. Unless both are present, the peculiarity is not produced.

It is clear from these cases that even though an individual inherits the type of gene necessary to produce a certain defect, it may not be inevitable that he should have that defect. A special environment may correct what inheritance leaves imperfect. These considerations apply to many things in man as well as in other organisms.

Another example of the fact that what is produced in development depends both on the genes and on the environment is seen in the production of giants in the fruit-fly. A modified gene, located in the X-chromosome, near its left end, causes the animals (if they have no other type of X-chromo- somes than this) to become giants; they grow to nearly twice the size of the ordinary fly. But this increase in size takes place only if the animals having the modified gene are well fed during a certain period of their early lives. If not well fed at this particular time in life, they grow no larger than the usual flies. The giant size requires to produce it a particular type of environment acting on a particular type of gene. If either condition is not fulfilled, giants are not produced.

Again, in plants, the difference between green plants, containing chlorophyll, and white ones, is in some cases due to a difference in genes. Plants with a certain kind of genes will remain white even though grown in the light. In other cases, the difference between white and green plants is not the result of a difference in genes, but of a difference in environment. Plants that live in the dark remain white, even though they contain genes that can produce chlorophyll, while similar plants grown in the light are green. That is, to produce chlorophyll, a certain type of genes is necessary, and also a certain type of environment.

Again, there are red and white varieties of primroses; the difference in programmes is inherited; it is due to gene diversities. But a certain variety produces red flowers when grown in a cool place, white flowers when grown in a warm, moist region, as in a greenhouse. The same diversity in programmes is now due to an environmental difference.

Complex interrelations of genetic and environmental action have been described by Emerson with relation to diverse colours in maize plants. In different types, the leaves, the husks, the flowers, and other parts, differ in programmes. Emerson describes and Figs six main types, under each of which are subtypes. These are the following:

1. Purple, with its subtype, weak purple.
2. Sun-red, with a subtype, weak sun-red.
3. Dilute purple.

4. Dilute sun-red.
5. Brown.
6. Green, with a number of subtypes.

The interrelations of genetic and environmental factors in these colours may be summarized as follows:

1. Each programmes and each subtype is constant and hereditary, when the different types are bred under like conditions.
2. Each of the colours depends on many different genes, like the eye programmes in Drosophila; they may be altered by changes in any of these genes. The different colours result from different combinations of the genes.
3. When the different types are crossed they give typical Mendelian inheritance: in some cases with simple ratios, in others with complex ratios.

   Thus far the conditions are those for typical hereditary characters; the differences are the result of gene differences, and they follow the distribution of the different types of genes in inheritance.

   But, also, the colours depend on the conditions to which the plants are subjected while developing. Change of certain conditions changes the colours, as does alteration of the genes. Some of the relations are the following:

4. The 'sun-red' type requires exposure to the sun for production of the red programmes. Parts protected from the sun are green. In the 'green types, on the other hand, subjection to the sun does not cause the production of red programmes. Thus the difference between red and green plants is due in some cases to diversity of genes, as when 'sun-red and 'green types are both grown in the sun. In other cases the difference between red and green plants is due to environmental difference, as when 'sun-red plants

grown in the sun are compared with those of the same type grown in the shade.

The plants of the 'purple' type do not depend on exposure to the sun for their programmes; they are purplish in programmes whether grown in sun or shade.

The degree of programmes produced by exposure to the sun depends on what genes are present. In the typical 'sun-red' variety, exposure to the sun causes a red programmes widely distributed on the plant. In the 'dilute sun-red', exposure to the sun under similar conditions causes only a little red, at the bottom and tips of the leaves, while in the 'green type no red appears.

The nature of the soil affects the programmes. 'Dilute sun-red' grown in poor soil and subjected to the sun has almost all parts red; in good soil, with the same relation to the sun, it has very little red. The 'green and 'brown' types are not influenced in their programmes by the nature of the soil.

5. Certain particular chemicals in the soil affect the programmes. Presence of much nitrogenous matter tends to suppress the red; if little is present the red programmes is marked.
6. Storage of carbohydrates in the leaves of the programmes- producing varieties causes increase of the red programmes. The amount of carbohydrate in the leaves may be increased by experimental procedures, as by removing the growing ears, or by bending a leaf so as to break the conducting vessels; the leaves thereupon become red in the parts with increased carbohydrate. But this effect is not produced in the 'green' types; they do not produce the red programmes even in presence of stored carbohydrates.

It appears clear, therefore, that in maize different genes cause the different varieties to have different metabolic processes, different chemical reactions, some of which cause the production of colours, while others do not. Also, diverse environmental conditions likewise cause changes in these metabolic processes, inducing programmes production in some cases.Whether programmes occurs, and its type and amount, depend on interaction of genetic and environmental conditions.

Another instructive case of the interaction of genes with environmental conditions is furnished by the conditions determining the number of facets in the compound eye of Drosophila. In the normal eye there are present about 800 facets. In certain of the flies a gene at locus 57 in the X-chromosome is so modified as to cause a great reduction in the number of fully developed facets. In this condition, known as Bar-eye, there are only about 80 fully developed facets: these form a 'bar' across the eye. This Bar-eye condition is dominant; the number of facets is reduced even though one of the two genes of the pair is normal. We may call the modified gene that causes the reduction in the number of facets the Bar gene.

The number of fully formed facets, it is found, depends on whether there is present but one or two of the Bar genes, and also on whether there is present in addition a normal gene. In the males (having of course but one X), the presence of the Bar gene reduces the number of normal facets to about 90. In the female, if both the X's contain a Bar gene, the number of facets is about 65. If the female is heterozygotic, so that but one of her X's contains the Bar gene, while the other is normal, the mean number of facets is about 358. Finally, in either male or female, if no Bar gene is present, the number of facets is about 800.

Later it was discovered that this Bar gene is sometimes modified in a different way, so as to cause a less marked reduction in the number of facets. This modification is known as infra-Bar. A female containing but one of these infra-Bar genes has about 716 facets; if she contains two she has about 348.

Again, by an extraordinary process, almost or quite unique in known genetics, in some cases two of the Bar genes get united by crossing-over into one X-chromosome. This condition is known as double Bar. When both of the chromosomes carry this double Bar, the number of facets is but about 25. If one carries double Bar, the other single Bar, the number of facets is about 36.

Thus the number of facets in the eye depends in many ways on the number and conditions of the genes present. This may be shown in the following diagram, representing the conditions in the two X (or Y) chromosomes of any individual, with the corresponding numbers of eye facets present. In this diagram

+ = the normal, unmodified X-chromosome Y = the Y-chromosome of the male B = the Bar gene iB = the infra-Bar gene

BB = the double Bar condition (two Bar genes in one chromosome)

Ghr. i: + iB B iB B iB B BB BB BB BB Chr.s: + + + iB Y B B + iB B BB Facets 800 716 358 348 98 73 65 45 42 36 25

All these various conditions are inherited as unit-difference, sex-linked characters; they are typical hereditary conditions.

But it is found further that the number of facets in most of these conditions depends also on the temperature to which the developing animals are subjected. The numbers given above are those observed by investigators working with fruit-flies cultivated at ordinary, not precisely controlled, temperatures. If different sets of flies are kept at different temperatures, it is found that high temperatures reduce the number of facets in the individuals that carry the Bar genes, though not in the normal individuals. Krafka found the following mean numbers of facets in the various gene conditions, at different temperatures:

Temperature, Degrees	Bar Females B-B	Bar Males B-Y	Double Bar Females BB-BB	Double Bar Males BB-Y
J5	214	270	5i	61
20	122	161	33	37
25	8l	121	25 ‘	28
300131	40	74	‘5	‘4

Thus the number of eye facets depends both on the genes and on the temperature. The number 61 may be produced either by a double Bar Male (BB-Y) at 15 degrees or by a Bar Female (B-B) at between 25 and 30 degrees. Changing either the genes or the temperature will change the number of facets.

Similar situations occur in vertebrates; one of these may be described briefly. The spotted salamander, Salamandra maculosa, is largely black in programmes, with spots or stripes of yellow. The amount of yellow differs in different cases, so that on the whole some specimens are lighter, some darker. In the lighter individuals the yellow spots tend to run together into rows. The different grades are heritable; certain stocks are regularly darker, others lighter. If darker and lighter individuals are mated, the two conditions show Mendelian inheritance, the darker condition being dominant all the Fi generation are darker, while in Fa there are three dark to one light. The difference between the two thus depends on a difference in one gene.

But it can be shown also that the extent of programmes depends on the environment. If the salamander is kept continuously for a long time (a year or two) on a light background, the yellow spots increase in size: they may run together into stripes or into large yellow areas B, so that the animals become much lighter. If, on the other hand, they are kept for long periods on a dark background the yellow areas decrease in size; the animals become on the whole darker. The extent of the yellow programmes thus depends both on the genes and on the environment; it can be altered by appropriately changing either.

In all these cases the following general principles apply:

1. What are really inherited (passed on from parent to offspring) are the genes; that is, certain materials that in certain combinations and under certain conditions give rise to certain definite characteristics.
2. With the same original genes, different environmental conditions may induce the production of diverse characteristics.
3. Also, with the same environmental conditions, different genes may induce the production of different characteristics.
4. The same difference in characteristics that is in one case produced by diversity of genes is in other cases produced by diversity of environment.

There can be little doubt that these general statements apply to many characteristics in man. Some combinations of human genes form an individual that is a much better culture medium for the bacteria of tuberculosis than are others. A person who gets such a combination of genes will develop tuberculosis, if he lives in a region in which the germs of that disease are abundant; while another person, with a different set of genes, will not be subject to tuberculosis, even though living under the same conditions. And the person with the genes that make him susceptible to tuberculosis will not have the disease if he prevents infection by the tubercle bacillus.

In relation to such susceptibility there are doubtless a great number of grades, dependent on just what combination of genes is present. Some are extremely susceptible, others less so but still prone to the disease, and so on, up to individuals whose gene combinations are such as to give them natural immunity to tuberculosis.

What then shall be answered when it is asked whether tuberculosis is hereditary or not? One can only say that some gene combinations predispose to it more than others, so that a hereditary factor is involved. But also there is a necessary environmental factor; it is the interrelation of the two sets of factors that determines whether the individual shall be afflicted with the disease.

Similar considerations apply to any other disease, acute or chronic. Certain combinations of genes are doubtless more readily attacked by plague, by smallpox, by typhoid, by pneumonia, than others, just as certain genetic constitutions yield more readily to extremes of temperature, to exposure to the elements, to unfit food. There is no affair of life in which the gene combination borne by the individual does not play a part. But the environment too plays a part, often an overwhelmingly important part.

The relative role played by genes on the one hand, and by environment on the other, is not the same for all characteristics. Differences in certain features of the organism are mainly due to differences in the genes of which they are composed. In other characteristics, differences are due mainly to environmental diversities. There are all sorts of intermediate conditions between these extremes. We may pass in review a number of different types of characters from this point of view.

Most of the diversities in obvious physical features in most animals are the result of differences in the genes, so that they are mainly matters of inheritance rather than of environment. In Drosophila the forms of the parts of the body, of the legs and wings, the distribution of bristles, the venation of the wings, the colours of body and eyes, are all changeable through the action of diverse genes, but diversities in these features due to environmental action are relatively rare. This is the case also for such characters in the vertebrates, including man. Most of the physical characters of human beingsýÿ form of their features, complexion, programmes of hair, form of limbs, and the like, as well as sexýÿare settled mainly while the child is carried in the mother's body, when the differences in environment are very slight. The diversities between individuals in these respects are almost entirelyýÿthough not quite entirelyýÿthe result of differences in the genes; they are matters mainly of inheritance. This is demonstrated by what we find in twins. The one-egg twins, having the same genes, are usually closely alike in their physical peculiarities, while the twins derived from different eggs, and so having different genes, are much more diverse with respect to such physical characteristics. At times, of course, one-egg twins become diverse through accidents or diseases which affect one and not the other. But physical diversities in twins due to genetic differences are much more frequent than those due to environmental differences.

In some organisms the gross physical features are readily altered by environmental diversities, so that differences between individuals resulting from the different conditions under which they have lived are common. Such is the case with the habit of growth in higher plants; trees and other plants grown under different conditions are very different in form. In some of the lower animals, such as the hydroids, a similar situation is found.

It is obvious that the details of behaviour are in a high degree under the influence of environmental conditions. The particular acts performed at given moments in any given organism are determined almost entirely by the conditions under which it finds itself. Yet organisms of different genetic constitution behave very differently under effectively the same conditions, and the general pattern of the behaviour of organisms may be largely determined by the genetic constitution.

GENES AND ENVIRONMENT IN RELATION TO MENTAL CHARACTERISTICS

The relative rdles of heredity and environment are of special interest in man, particularly in relation to the characteristics that influence behaviour. For the study of these matters man is the most favourable organism that exists, although for most other relations in genetics he is a singularly unfavourable object of study. In man it is possible to make detailed studies of temperament, mentality, character, a group of characteristics hardly open to examination in other organisms. Furthermore, there occur in man 'identical' or one-egg twins, individuals having the same set of genes, the same genetic constitution throughout. And for comparison with these there occur fraternal twins: two individuals of the same parentage, the same age and living under the same environment, but developed from two separately formed eggs and thus having to some extent different sets of genes. By comparing these two kinds of twins with one another and with unrelated individuals, opportunity is presented for comparisons of the results of similarities and diversities in genetic constitutions with those of similarities and diversities in environmental influences.

As seen earlier the twins derived from a single egg are as a rule very closely alike in most physical respects; it is this that has caused them to be called identical twins. In some cases, however, the two differ in certain physical characteristics. There appear to be two categories of causes for these differences. First, there are physical differences that result from the way in which the separation of the twins occurred. Separation of the single egg into two does not occur in the first stages of development, as by division of the one-cell stage into two. On the contrary, it occurs much later, when many cells are present and the differentiation of bodily parts has begun, and it occurs at different ages in different cases. There is reason to believe that when an equal division occurs at a rather early stage, before the right and left halves of the body have begun to show their characteristic diversity, the two twins produced are closely alike in their physical features. At a later period the right and left halves of the embryo have begun to differ, in the way that results as a rule in the greater strength and development of the right hand and arm. At such a stage, the embryo may divide into right and left halves, with results as follows. Of the two embryos so produced, one will have a right limb that is already somewhat advanced in development, while the left limb, formed at the plane of separation, must begin anew. The other embryo will be in the reverse situation; it will have a left limb already somewhat advanced in development, but the right limb, formed at the plane of separation, must begin anew. In each case the more advanced limb retains its advantage and becomes the one that after birth is strongest and most used. The result necessarily is that one of the twins remains right-handed, while the other becomes left-handed. This situation is not uncommonly found in identical twins. Similar differences in other unsymmetrical parts may be produced in the same way: for example in the whorls of hair on the head. In such respects the two members of a pair often show 'mirror-imaging', presumably due in each case to the fact that the original embryo had already become a little unsymmetrical before division occurred.

From this same situation another condition often observed is believed to result. In the embryo before division, one side (usually the right) will be a little in advance of the other. After division, the twin derived from the right half continues to retain this advantage, so that it may be more vigorous than the one derived from the left half. Such differences in vigour are not rare in twins. Apparently they may be accompanied by a psychological difference; the more vigorous twin assumes the leadership in their lives together. This reacts further on temperament and character. How far psychological and temperamental differences so produced may go is uncertain.

It appears that in some cases division of an embryo may be unequal, so that for this reason also one twin may be more vigorous than the other, with resulting psychological differences.

Diversities produced in the ways above described are different in origin and type from those commonly called environmental. Under the latter term are commonly grouped diversities resulting from the different conditions under which individuals have lived. Since the extent to which differences due to the time and method of division may go, and the kind of characteristics that they may affect are uncertain, it is often not possible to distinguish these certainly from diversities due to differences of environment after division has occurred.

Belonging to one or the other class, there are sometimes marked differences in the phsyical characteristics of one-egg twins. Komai and Fukuoka have described two fifteen-year-old Japanese boys that are obviously one-egg twins; they show the usual close resemblance in form, features, colouring, and the like. But they now differ greatly in size. One is taller by 14-8 centimetres, or about 6 inches, and his weight is greater by 10-4 kilograms, or about 23 pounds. The two were at first of the same size, the difference gradually arising during growth. The smaller of the twins is affected by diabetes insipidus; this is probably connected with his decreased growth. The larger twin is right-handed, the smaller left-handed. Whether the difference in health and growth is in some way connected with what occurred at division of the egg is uncertain.

A comparable case is described by Siemens. Two sisters, clearly one-egg twins, were as usual nearly identical in most of their physical characteristics till the age of 10. At this time one of them became affected wth a severe lateral curvature of the spine, and from this time their development was very different; at 16 the healthy twin was 4-8 inches taller than the other. The origin of the defect that produced curvature of the spine and changed development is unknown.

Identical twins that have lived together sometimes show considerable psychological differences. Newman has made extensive psychological tests on fifty pairs of identical or one-egg twins, and, for comparison with these, on fifty pairs of fraternal twins. The results of these examinations have not yet been published in full, but certain data from them have been published. By the Stanford-Binet tests, the intelligence quotients (IQJ and the 'mental age' were determined for the two sets. The two classes of twins are best compared by the differences between the scores made by the two members of each pair. These differences were as follows:

Mean difference Mean difference in Mental in Intelligence Age Quotientmo. 5-3 ponts

'5 years 9 mo. 9- 9 points

Thus the average difference between fraternal twins, having somewhat diverse genetic constitutions, is nearly twice as great as that between identical twins, in which the genetic constitutions are alike.

But it is important that the identical twins that have lived together also showed differences between members of the pairs. In different pairs there was much variation in the amount of diversity. Of the 50 pairs there were five in which the two members gave identical scores in the intelligence tests; the difference of their scores was o. At the other end of the scale were five pairs that showed respectively the following differences (points in the Stanford-Binet scale) between the two members: 12-6, 12-9, 13-0, 13-9, 16-0.

All these five differences are considerably greater than the average difference between the genetically unlike fraternal twins. It is clear that in some cases even identical twins living together differ much in mentality. Whether the differences are the result of original diversities consequent on the method of division of the egg, or have been produced in some way by different experiences of the two, is not known.

For comparisons as to the relative effects of genetic and environmental diversities on mental characteristics, valuable data have been obtained from the examination of identical twins that have lived apart under different environments. The first case of this kind was carefully studied by Muller, and nine pairs have since been fully examined by Newman. In each of these ten cases the twins were separated in infancy, being adopted into different families; they have then lived to adult life under diverse conditions and influences. In addition to careful observations as to character and temperament, they were subjected to psychological and temperamental tests by the best standardized methods available.

The type of results reached by these studies will best be appreciated by first looking at the detailed resemblances and differences in some particular cases. In the first such study, by Muller, there were twin sisters that had been separated when two weeks old, and that did not see each other till they reached the age of 18; from that time until the age of 30 they lived apart more than nine-tenths of the time. Physically they showed the extreme similarity in characteristics that is usual in identical twins. Both 'have always been intellectually active', 'both have been extremely energetic, capable and popular, and they have been prominent in all sorts of club work in their respective communities' (Muller). 'Both have had two or three attacks of tuberculosis, almost simultaneously'. The usual intelligence tests gave results very closely alike for the two twins. But 'the nonintellectual testsýÿof motor reaction time, association time, "will temperament", emotions and social attitudesýÿgave results in striking contrast with those of the intelligence tests, in that the twins gave markedly different scores in all these tests'. The differences were on the average greater than those between two individuals taken at random, and seemed 'to be correlated with salient differences in their past experiences and habits of life'.

Thus this first study of such a case indicated that the different environments and experiences of the two individuals had produced a large effect on temperament, emotions, and social attitudes, but had had little effect on such matters as are tested by intelligence tests.

An illuminating contrast with these results is given by the first pair studied by Newman. The twin sisters ('O' and 'A') were born in London and were separated at the age of 18 months. One lived in Ontario, Canada, the other in London.

Their environments were very different. When they were tested, this pair gave differences in those tests in which Muller's pair gave similarities, and gave similarities in those tests in which Muller's pair gave differences. Newman says: 'The twins dealt with in this paper are very different in mental capacity. But they showed great similarity in their manifestations of will and temperament, and in their emotional reactions. So this pair of twins shows that differences in the experiences undergone affect deeply the individual's mental traits: his performance in matters brought out by intelligence tests.

Important points were brought out in the study of Newman's second pair, twin sisters ('£' and 'G'). The two had received very different educations; one had attended school seven years longer than the other. Newman summarizes as follows the results of his study of these:

'These twins, remarkably similar after being separated at 18 months of age and unknown to each other for 19 years, have been profoundly modified by the very different educational careers.In every test of mental capacity, whether of so-called native ability or of achievement, 'G', the more highly educated twin, has distinctly the superior mind. Obviously mental training improves the ability of an individual to score well in any sort of test'. But further, 'in contrast with the great difference in mental power stands the fact that in all the tests of emotional traits and of temperament the twins gave the impression of being remarkably and unusually similar'. Here came out the effect of their identity of genes.

Newman's third pair revealed certain other facts of importance. These were two young men ('G' and 'O'), separated at two months of age. One had lived mainly in the city while the other had lived in the country. They were examined after reaching the age of 23. The results are summarized by Newman as follows:

'In native ability they seem to be nearly identical. The one outstanding difference is in their general personalities. 'G' (who had lived in the city) impresses one as more dignified, more reserved, more self-contained, more unafraid, more experienced, and less friendly. He seldom smiles, has a more serious expression about the brows, eyes and mouth. He stands more erectly with chin held in and brows drawn down somewhat over his eyes. 'O' (the other twin) is the opposite in all these respects. He is the more typical country boy, laughs readily, and is not on his dignity at all.' Newman asserts emphatically that 'the personalities of the boys were utterly different'.

Thus through the study of these first four pairs examined, of identical twins that have lived apart, it became clear that such twins may differ in many ways. It appeared that the different environments and experiences of the individuals may have a large effect on mental and temperamental characteristics. The effects were different in ctifferent cases: in some cases the twins are alike in intelligence but differ temperamentally; in other cases they are alike in temperament and emotions, but differ in mentality. In other cases the marked difference is best expressed rather vaguely as a diversity in 'personality'.

The general impression given-by study of these first four cases has been confirmed by Newman's study of six other pairs of identical twins that have lived apart. The nine cases examined by Newman were treated by uniform methods, so that they may be compared in such a way as to bring out general relations.

The intelligence quotients for the two members of the nine pairs are as follows (Stanford-Binet scores in points):

Pair IQ, Difference
I 97 and 85 12
II 67 and 78 ii
III 99 and 101 2
IV 106 and 88 18 V 93 and 89 4
VI 102 and 94 8
VII 106 and 105 i
VIII 92 and 77 15
IX 102 and 96 6

The mean difference in intelligence quotient for these nine pairs of identical twins that have lived apart is 8'6. This is considerably greater than the mean for identical twins that have lived together, which was 5-3. So far as it goes it indicates that difference of environment has a considerable effect in increasing the mental difference between twins. In cases I, II, IV and VIII, the difference between the twins was very marked, but is similar to the differences between the five extreme cases of identical twins that had lived together. This emphasizes the fact that it is not possible to be certain what differences are due to something that happened at the division of the egg, and what are due to later environmental differences. The.mean difference between identical twins that have lived apart (8*6) is nearly the same as that for fraternal twins that have lived together.

In addition to the studies of intelligence, Muller and Newman made extensive examinations designed to test other features of personality. One set was designed to test will and temperament, to distinguish the mentally quick and slow, the deliberate and careful as compared with the hasty and careless, the aggressive and forceful or the opposite, and the like. Another type of examination is designed to test emotional peculiarities: likes and dislikes, feelings of right and wrong, and so on. The results of such tests are not readily expressible by a numerical score, such as is employed for the intelligence tests. It will be worth while, however, to attempt a tabulation in general terms of the results of the tests in the three categories for the 10 cases of identical twins that have lived apart. The first case (M) is that of Muller; the others are the nine cases of Newman:

Intelligence Will-Temperament Emotions
M Closely alike Very different Different
NI Very diverse Alike Alike'
II Very diverse Closely alike Closely alike
III Closely alike Closely alike Very different

IV Extremely diverse Very different Rather alike V Closely alike Closely alike Closely alike Intelligence Will-Temperament Emotions VI Rather alike Rather alike Rather alike VII Closely alike Somewhat differ- Somewhat different ent VIII Very different Considerably dif- Somewhat differferent IX Much alike Closely alike Considerably different

Thus the general picture is that identical twins that have lived apart show many different combinations of likeness and unlikeness with respect to intelligence, temperament and emotional characteristics. There are cases in which the two are closely alike in all three categories (V and VI), cases in which the twins are distinctly diverse in all three categories (VIII), and cases in which they are alike in two categories and diverse in the third, or alike in one category and diverse in the other two.

On the whole, it is clear that other things, beside genetic constitution, play important roles in determining mentality, temperament and emotional traits. The principal difficulty in interpreting the results lies in the uncertainty as to how much is the result of the manner in which division of the egg occurred, and how much is due to later environmental differences. Both certainly play a role. For these types of characteristics it appears probable that, of the two, the later environmental differences play the greater role. It is further clear that genetic resemblances and differences play a very large role in mentality, temperament and emotional traits.

It is to be noted that in all this study, the general environment of the twins was much alike in all cases. None of them lived in different civilizations, or at different epochs in cultural development. How great a diversity in the characteristics studied could be made by such differences remains uncertain.

It is further to be noted that the tests to which the twins were subjected were designed to bring out rather permanent personal traits, as distinguished from mental content and habitudes. The behaviour of human beings depends very largely upon acquired mental content, upon knowledge, and upon habitudes, and these things are in a high degree dependent on the conditions under which the individuals live and the experiences to which they are subjected.

In general, it is clear that in a human population there are great numbers of different types of individuals that are diverse because their genetic constitutions are diverse. In consequence of this they differ in their capabilities, in their tastes, in their tendencies toward any particular line of action. But all these classes, so far as they do not fall in the seriously defective groups, have marked powers of adjusting themselves to many different conditions and of following many different lines of action, depending on the conditions in which they live.

In no other organism than man is so full an analysis possible of the relative roles of genetic constitution and environment in determining the mental and temperamental characteristics, for in no other organism are these characteristics highly developed.